Beeing

Also by
Rosanne Daryl Thomas:

Novels

The Angel Carver

Awaiting Grace

Humor

Complications (by Prince Charming)

Coffee: The Bean of My Existence
(written and illustrated by R. D. Thomas)

Beeing

Life, Motherhood, and
180,000 Honeybees

Rosanne Daryl Thomas

The Lyons Press
Guilford, Connecticut
AN IMPRINT OF THE GLOBE PEQUOT PRESS

The Lyons Press is an imprint of The Globe Pequot Press.

Printed in the United States of America

10 9 8 7 6 5 4 3 2

Text design by M. A. Dubé

Library of Congress Cataloging-in-Publication data

Thomas, Rosanne Daryl.
  Beeing : life, motherhood, and 180,000 honeybees / Rosanne Daryl
Thomas.
    p. cm.
Includes bibliographical references (p.   ).
  ISBN 1-58574-731-9 (hardcover : alk. paper)
  1. Honeybee--New England--Anecdotes. 2. Bee culture--New
England--Anecdotes. 3. Thomas, Rosanne Daryl. 4. Beekeepers--New
England--Biography. I. Title.
  SF523.3 .T56 2002
  638'.1'0974--dc21
                                            2002013071

For
my daughter, August,
and
Connie Kaiserman,
with love

and for everyone who knows
that bees don't bite

The sting of the bee
and the dart of Cupid
both signify the secret fire,
the mercurial solvent which
destroys the old metal or
outmoded state of being.

—*A Dictionary of*
Alchemical Imagery

Contents

Prologue

ORE THAN ONCE HAVE I SEEN STARTLING photographs of men wearing bee beards. What the man does is take the queen from her hive and put it on his face. Then, bees, being the way they are and ever will be, hurry to surround their queen, clinging to each other, clustering, until whoever has seen fit to move her creates an animate approximation of biblical facial hair. This is the sort of thing that people tell you their uncles did at Vermont county fairs. It is, as far as I can tell, what is known as A Guy Thing. I see neither the point nor the allure. Beekeeping is rather A Guy Thing too, though a seven-thousand-year-old cave painting in the Cuevas de la Araña of the Spanish Levant shows what might well be a woman in the act of gathering honey. The gatherer, who appears as a naked silhouette with an hourglass figure, climbs up a cliff to reach her bare hand into a cranny filled with bees and pulls out sweet, dripping combs. Though some historians claim this shapely gatherer is a man, I think not. What I wonder about is how it was she climbed the height and who was waiting at the bottom. Her children? A man?

Like so many passions, my relationship with bees and beekeeping was unsought, unlikely. Shall I say it was unavoidable? No. But once I started, there was no stopping.

Before I had ever seen a hive up close, I glanced at a beekeeping manual with vague curiosity. I remember the warning: "If you want honey, you have to be prepared to get stung." I thought, That about sums it up—"it" being everything.

The sweetness of life does not come without the occasional nasty sting, but those who go too far out of their way to avoid the venom avoid the sweetness as well. When my life was sweet I had always savored it entirely, enjoying every drop. But when it was painful, I had to learn and relearn what life and beekeeping had to teach. If you want to get honey, you have to be prepared to get stung.

1

To Just Be

The hive is sweet
because of the honey in it,
but you must be sweeter
as you are the source and spring
of the honey.

—*A hornet to a bee, from the poet Nazimi*

I ALWAYS BELIEVED THAT IF I WERE TO MARRY, I would marry for life. I knew it was possible. My grandparents were married for fifty-one years and in love for longer than that. One night my grandmother died, swiftly and quietly. When my grandfather found her he cried out, "No! You can't go! I still have so much I want to tell you!"

Theirs was my favorite love story. When I married, I wanted the kind of love they had had, fifty-one years of so much to tell and so much to learn. But I chose unwisely. Instead, I had five years of "working on the relationship" followed by a hideous divorce. But I will spare us both the gruesome details. This is about what came after, and how I ended up with the bees.

The way through dark times is never easy. The stones are sharp under your feet and your heart feels unprotected by your ribs. I needed to calm myself so that I could prepare for whatever—and who knew what "whatever" might be—needed doing. My head was host to a tornado of worry and distress. Half the time I couldn't eat, the other half, I'd forget to. After a long day strangely mixed with blessings and fear, I'd fall asleep beside my little daughter, storybook in hand. Then I'd wake, find my way to my own bed, wake again at 2:30 in the morning to the loud noise of my thoughts and then again at four, before the sun rose and the birds started their day.

On the advice of a friend, I took up a beginner's form of Tibetan Buddhist meditation. It helped. At least I remembered to breathe, which is not always as automatic as it sounds, and I tried to remind myself that while I could not amend the past, know where next to set my foot or whether to answer the phone, much less what the result of all this harsh tumult was going to be, I could lay claim to the present, where moment to moment I was well enough, safe enough, even happy enough to find some kind of peace and move on to the next present moment. The phrase used to describe this gentle relationship between thought and time was "to be," more exactly, "to just be."

During that time, I found a postcard of a fuzzy yellow and black bee standing on five of its six legs. It was all alone on a big white background. One wing looked a little raggedy. I took the bee home with me and put it in a frame on my desk so that each time I looked up I saw it and thought, Just be.

My daughter and I spent one year in a cottage on a farm known for endless vistas and boundless fields of daffodils—a paradise. But our paradise came at a price we could no longer afford to pay if we meant to keep up the habit of eating three meals a day. So we moved to the center of a small New England town where the noon whistle blows at 12:06 and the population runs about three thousand. We rented the top floor of a two-family house. It was small, dingy, and rickety. Colorful paint would fix the dinginess. The rest we could ignore. There was enough space for the two of us, our old gray cat, and my laptop. Room to begin my new novel and something to stare at while I daydreamed about the lives of people who existed only in my imagination. That was most important. Our second-story flat had a balcony from which we could see hills and sky and river. My daughter and I need to be able to see those things. That's the way we are. Our kitchen window looked into the branches of a dogwood tree where we watched the birds from a bird's-eye view. What we saw was always changing. My daughter called that window "the magic curtain."

When we moved away from our cottage on the farm, my daughter had had to leave behind a wild strawberry patch. We might not yet have friends in our new town, but we could have strawberries. We planted delicate French *frais du bois* in pots set out on the balcony railing so they could make the most of the sun. My daughter wanted a blueberry bush too. And why not?

That summer, we had a few strawberries and the blueberry bush bore fruit, but reluctantly, so that each berry's arrival was an event, and we tried to time the picking to the ripest moment before our tiny harvest would be snatched by the squirrels and birds.

The next summer, our blueberry bush was sad to behold. Though we cared for it as best we could, it had nothing to give. We went to the greenhouse that sold us the plant and asked why this was so. The lady at the greenhouse knew the answer. We had the right soil and we had the right sun and we'd done the right things, but though we had birds and squirrels, mosquitoes and little black flies, what we lacked sealed the bush's fate. There was no hope for it. It had not been pollinated for two summers. "All it would have taken was one little bee," said the greenhouse lady, and my life took an unexpected turn.

2

Spring Ahead

He must be a stupid countryman
indeed who cannot make a beehive:
and a lazy one indeed, if
he will not if he can.

—Mackenzie's 5000 Receipts, 1829

*T*O SAY THAT I THEN BECAME INTERESTED IN beekeeping would be nothing short of a lie. It may be, however, that at that moment beekeeping became interested in me. I am the last person anyone would call a nature girl, but I do not deny that I knew something about bees. They were yellow. They buzzed. They stung. And somehow they made honey which got put in teddy-bear jars. They interested me less than the stars, the sight of which brings me ignorant delight; and more than cream-based sauces, which I avoid.

There you have it.

But once something decides to become a part of you, it finds a way. I went to an outdoor wedding. An elegant woman wearing a bejeweled bee brooch was visited by the brooch's living inspiration. The woman forsook her elegance and flapped about in horror and dismay. I wasn't surprised, but, for the first time, I realized just how peculiar it was that she ornamented herself with the false bee and despised the real one. Hating insects amounts to hating most of the population of the world. That same afternoon, the bride's sister told me that despite a severe allergy, she believed she had a special, almost mystical, relationship with bees and so, despite the danger, she did not fear them. Bees buzzed about the mouths of empty beer bottles and an erudite man wearing a bow tie quoted from memory both J. H. Fabre and Maurice Maeterlinck on the subject of bees. Bees were in the air and in the air.

When I returned home, I called the local out-of-print book dealer in search of Fabre and Maeterlinck. The volumes arrived; handsome old books with gold-embossed covers that formed the foundation of a new to-be-read pile on my bedside table where they sat along with forty other books on assorted subjects that wait, sometimes indefinitely, in untidy pyramids for my late night attentions.

Three seasons passed. Maybe more. I had fallen in love with the work of a local sculptor who possesses the gift of giving life to the animals he casts in metal. It is as if he has some deep and simple knowledge of their spirits, the essence of which he embeds in each piece. I'd heard that, in his wife's store, something, not quite what I yearn for, but something very very small, a tiny crow perched on a black bronze teacup, might be had for under two hundred dollars. So I found this store. All the tiny crows were gone but the sculptor's wife most graciously, and fatefully, offered me an invitation to visit his studio.

March 9th was the day. It was wet and rather icy. I'd worn the wrong shoes for my pilgrimage. I hadn't known we'd be hiking up a small mountain but that is what we did, I, my almost seven-year-old daughter, and maybe fifty others, the ardent and the curious. During the trudge, my daughter declared that we were ascending Mount Olympus. Being half adult-size, the hill was twice as high and very imposing. Sometimes everything comes down to scale.

As we reached the top, we entered the kingdom of silent animals. On the rocks, and in the lawn, the rats, the cats, the crows, the goats, the bulls, a bronze barnyard, stood still; as if while a cloud blew overhead, time stopped and would, when it passed, begin again. Inside the studio were four tremendous white plaster and foam elephants. They stood like guardians of the four winds, silent trunks raised above our heads, their eyes too high to meet a human gaze. They were about to board a slow boat to China. When they reached Shanghai, these awesome white creatures would be cast, in a single pour, into the three-ton bronzes they were destined to be.

I stared, my imagination full of sparks and molten metal, white elephants turned to bronze. Someone shattered my daydream with his voice. The speaker was a lean blond man with a wide, open face and a tidy, old-fashioned moustache. He said he was an organic farmer. He called himself Farmer Tom and gave me his card. Feeling a twist of skepticism, I put it in my purse without looking at it. I'd never met a farmer with a business card, or at least not so I'd known it, and he looked altogether too neat for a man whose hands worked the soil. But then again, it was early March, too soon for planting, and there was no reason to come muddy to the elephants' farewell. I thought he was rather nice to look at, so I asked him what he grew. He named some things I'd heard of and some I hadn't. Then, with sudden excitement, his blue eyes brightened and he told me he had a plan. Not this year, but maybe next, he was hoping to procure and grow some exotic Eastern European berries that, when crushed and sweetened with honey, made an unusual and unforgettable drink, a nectar worthy of the gods. I, without giving any weight to my words, found myself remarking that someday I might like to keep bees, and if I did, I supposed I'd have honey. My little daughter smiled at me. She liked the idea of bees too.

"Why not start now?" asked Farmer Tom. "You can keep them on my land."

So there it was. Simple as that. Only not.

For the next day and even the next after that, my daughter and I imagined ourselves as beekeepers. She danced jubilantly about the house, composing beekeeping songs, drawing beekeeping pictures. She drew me standing beside a hive that looked just like the ones we'd seen at a distance here and there on back country roads. Underneath the picture, in her very best handwriting she'd written, "MY MOMMY is a BEEKEEPER!!!!!!" I counted the exclamation points and thought, Uh oh. What is going on?

Standing under a herd of plaster elephants, I'd made a remark, a weightless, chit-chatty remark while conversing with a stranger, and now it seemed that my idle, unconsidered fragment of a fantasy was taking a fuller shape. Had I intended this? Certainly not. Maybe someday, some unthought-of, undetermined day. Not now. But my pretty metaphor had seized its moment and spun a silk cocoon. It was metamorphosing at warp speed into a real thing. While I'd seen the reverse many times, from real to a figure of speech, this was a rarity. I realized that if I didn't stop it, my metaphor would have a life of its own. I would soon be more than "just being"; I would be a beekeeper, which, at the very least of it, involved bees, about which I knew a teaspoonful more than absolutely nothing. I was afraid, but more than that, I was curious.

I checked my bee books. According to the assembled wisdom, a hive in a coolish climate should be placed on as treeless a hill as possible, preferably sloping south or east near a fresh water source with the entrance facing southeast so that the bees will feel the early sun and get cracking right away instead of lazing about the hive reading the newspaper. I gathered that the hives should have a bit of shade, not too little nor too much, in the heat of the day, though how they were to have any shade on a tree-less hill rather puzzled me. Then, of course, come winter they should be exposed to as much warming sun as possible while also being sheltered from the cold north wind. It all sounded very complex.

What if all the beehive-ish spots on Farmer Tom's land were facing the wrong direction? The only way I know right from left is that one hand has rings and the other picks up the pen. I would need the old compass that sat in permanent west-northwest on the kitchen windowsill to distinguish any directions other than up and down. And, starting to see the fun in this, I realized that we already had the perfect footwear. My daughter could wear her green froggy-face boots, brought for her all the way from England. I could find an excuse, after years of waiting for the right occasion, to strut about in my catalogue-purchased perfect purple Wellington boots with the flappy buckle and strap at the top. Well shod, we could at least pretend to pace soggy fields in search of proper places.

If only I could find Farmer Tom's business card. Or remember his full name. I could do neither. His card wasn't in the kitchen between the spice rack and the change jar. It wasn't in my bedroom, my pockets, or my purse. It wasn't in either of the shrines to fleeting contact where I drop the scribbled addresses and printed cards of people, restaurants, shops, "just in case." It was nowhere. Exactly where I put it. My psyche was clearly not willing to humor me. I took revenge on my psyche by deciding not to worry about it.

Every week or so we had occasion to drive south along a river road that starts out lined with trees and flowing purple loosestrife and gradually grows ugly, taking us past Kmart and Super Stop & Shop. Every week, twenty-three miles down, just before the road turns irreversibly suburban, we passed a little bunker with a sign that read "Beekeeping Supplies." This time, we did not pass. "We'll ask a few questions," I told my daughter, though in truth I was so unfamiliar with the subject that I was not up to questions yet. This did not prevent the proprietor, a master beekeeper with three hundred hives of his own and a pleasant manly face, from having answers. His eyes were soft and inquisitive. He seemed to be evaluating me, but without reaching a conclusion: "What do you want to know?"

"I'm not sure," I answered truthfully.

My answer suited him. I could see he preferred honest ignorance to half-knowledge. He pointed to a wooden box with a handhold and a hole. It was about nine inches wide and roughly my personal cubit—the distance from finger to elbow—long; the size of a Christmas package containing something better than a shirt. This hollow, topless and bottomless box was called the hive body or the deep super. This was where the bees lived. And it would be filled with ten wax-covered moveable wooden frames that resembled oversized doll-house windows with yellow shades and dangled on an inner rim, lined up inside edge to edge. It was on these unimpressive toy window frames filled with wax that the bees would make their honey. The super was light now, and the demonstration frames didn't look hearty enough to withstand a breeze, much less the making of honey, but he promised one box all together would weigh ninety pounds when the bees had been at their labors all summer. It did not come ready-made. Neither did the wooden frames. They had to be assembled and the wax, which came in sheets, fitted in. And then there was the Tin Man's head, with a tapered spout growing out of its forehead and a yellow bellows attached at the back. That was the smoker and you had to know the right way to light it, but I was not to worry about that yet, which was good because I wasn't ready to worry. There was a wonderful shiny silver-colored tool that he claimed, because of its brightness and shininess, would never get lost even in tall grass. I immediately liked this tool. It was about nine inches long, straight, mostly flat, had no moving parts, and did not need assembly. This was conveniently called the hive tool, and he claimed it was all a beekeeper would need, toolwise.

"How tall are you?" asked the Bee Master. It was obvious he already knew. Was he testing me?

"Five-three."

"Kind of small," he commented. He retreated to the back room, which was full of shelves.

He returned with The Outfit. I am extremely vulnerable to the allure of an Outfit. In my closet hangs everything I need to be a stylish fly fisher. Even though I live on (not near, but on) one of New England's famous fishing rivers, even though I know how to tie my own flies and vaguely remember how to cast them; even though my old fishing buddy sends me an engraved fishing trophy every year, a leaping bronze trout on a plastic base; I have not set my waders in the water for three years. The trophy is meant to shame me, in an amiable way. The base is always engraved with a reprimand: CONSERVATION AWARD. I plan to put on my waders and get back in the water one of these days. Meanwhile, I have The Outfit and I love it. I can look at it hanging smartly in my closet—next to the neglected evening Outfits, so rich with possibility in velvet, silk, and taffeta—whenever I want. Having the garb is not getting out on a stream, but it is still distinctly satisfying even as it points out the gap between doing and daydreaming. It's ready to go whenever I am.

First, the Bee Master handed me the gloves. "Try 'em on," he said.

How could I not? Elbow-length unbleached cotton to the wrist and from that point to the fingertips, sheathed in sunshine-yellow latex. I wiggled my fingers and wondered if they might come in a 7½. I dared not ask. "You wear this when you're working in the hives?"

"Some people do. If you're scared of getting stung. But you probably won't."

"I won't?"

"Nah. You don't look like the timid type." If I was, I wasn't going to admit it. "But you ought to have 'em anyway," he added. "In case." He dropped a white packet sealed in plastic onto the counter. He pulled off the wrapping and shook out a pure white beesuit. It looked like something you might use to repair the inside of a space station. The smallest size was a men's 42 regular, rather immense. Still, it was cotton. With a little improper care, it would surely shrink. The beesuit had a half-zipper yoke from shoulder to shoulder, front to back. The other half of the zipper was attached to yellow netting. The netting was fine and sturdy with even finer black wire mesh panels where the face would be. Though zipping the netting to the beesuit would seal you in and bees out, the netting had a hole where the top of your head would be. And what was the point of that? It fit over your hat. I was hoping for something in a pale gold closely woven straw. Something broad-brimmed and, perhaps, decked with an ornamental silk flower. He handed me a hard white plastic pith helmet with ventilation grates at the temples. Though I found the rest of The Outfit appealing, and was already debating whether the *soigné* purple wellies would be right, the hat was completely wrong. Something would have to be done about it.

"Where are the bees?" asked my little one impatiently.

"They come on a truck. You order 'em today. They'll arrive May second, give or take." Did I dare?

My daughter turned to me. "I could get my own beesuit and help you," she said.

In my time and culture, Socrates's wise maxim has been taken to extreme. Life is too often overexamined and underlived. I have been nose to nose with death twice and I know the sword will one day drop, so I have done a little rearranging. For me, the unlived life is not worth examining. But that is a grand motto and hard to live up to. When I, as a child, heard adults croon the lament of lost opportunity—could've, would've, should've—it irritated me to no end. I would almost always think, Well then, why didn't you? And sometimes I would ask. There was always a reason. Often the reason was an excuse. I knew the difference. I hated excuses for life unlived with the fierceness of inexperience. I swore I would not say "could've, would've, should've" to my children. I would live without regret. My daughter's wide hazel eyes were watching me. Here she was, newly seven, and ready to suit up and come with me to the buzzing, stinging bees. Was I going to find an excuse? Ashamed of my own lurking cowardice, I pulled out my MasterCard.

Six hundred dollars later, I became the owner of not one, but three yet-to-be-assembled hives, along with their bottom boards, outer covers, inner covers, a smoker I did not know how to work, a hive tool, a beesuit in need of shrinkage, gloves, netting, a skinny textbook, and a ghastly hat. As the smiling Bee Master helped me load his merchandise into the trunk of my blue Honda, he offered this: "Did you know," he said, "that the Catholic Church uses beeswax candles because beeswax is made by worker bees and worker bees are virgins?"

To which my daughter replied, "What's a virgin?"

We were starting at the very beginning. The date was March 28. In just over a month, six living pounds of Italian honeybees (Italian because the Bee Master said Italians have the best disposition, which, if you have ever been to Italy makes a lot of sense) and three young virgin queen bees, all yet-unborn larvae being nurtured by their older sisters somewhere in the state of Georgia, would be mine to tend.

Though the imminent arrival of approximately thirty-six thousand honeybees ought to have propelled me into action, it was all I could do to open the trunk of my little blue Honda and stare at the contents. The very sight of those brown paper packages filled me with a sense of accomplishment, as if, by the very purchase of these goods, I had been infused with what my daughter calls a "ness." In this case, beekeeper-ness. But, when, after a few days of checking my trunk and savoring my ness, the items therein did not assemble themselves, I began to suspect that there might be something more I ought to be doing. My mind became muddied and my ness was obscured by the paralysis of a new-found feeling. Dread.

Meanwhile, I was beginning to wonder what I was to do if Farmer Tom's business card chose to stay lost. I half-hoped for a graceful escape. Fate laughed. Farmer Tom tracked me down. The man had pollination on his mind. He was nice enough, but nice is not enough, at least not for me. Despite his wide, honest face, I was not in a hurry to trust him. It turned out his farm was not exactly a farm, nor was it exactly on his land. For a modest fee, he had rented eleven acres from the wealthy Land Trust people whose stated goal was "preserving the rural character of the town." Rural character is one of those things that resides in the eyes of the beholder. The local Land Trust did not choose a rural character to preserve the rural character. They didn't want a "grubby local" who knew his way around the dirt. They chose to create a farm, an organic farm, run by a socially deft, clean-handed preppy organic farmer of unknown expertise who hailed from a nearby community where gardens are cultivated by professional gardeners and not by the people who own them.

I slid on my wellies and drove south down the river road to the spot, which lay on the riverbank. The neatly shorn field was still waiting for the workings of spring to become apparent and so it wore the mantle of gentle gold that had seen it through the winter. At the edge, the river ran so peacefully as to seem entirely still, for it too was active only in unseen ways. It whispered, though. The water could be heard. Across the river I could see the rocky hills, naked but for the trees waiting to bloom. Near the road's edge there were two New England barns, weathered and worn, but stately nonetheless. Thirty steps from the barns there was an inlet, a shallow pool fed by the river, yet apart from it, well appointed with rocks and fallen logs. Birds rested in the bright and quiet water. Birds leapt up from the pale grasses and surveyed the waiting land. Birds flew from the barns and back again. Song was everywhere and the sun shone on all of this so that the air was warm and sweet.

I returned to my home feeling grateful for the chance to come to that field and know it, to walk across the land under the old hills by the river instead of driving past it every so often with a cursory glance at the view. At last I unloaded my trunk.

I had saved up my money so that I could get my grandmother's Empire loveseat fixed because, as it was, it was so frail it could only support the weight of a cat. In order to do the fixing, it had to be taken apart down to the ancient horsehair, which meant it had to be reupholstered, which meant that in my bedroom there was a hole between two windows where the couch used to stand. This hole was rapidly filled by three white boxes about two inches high with silvery galvanized aluminum tops, three boxes containing wood with notched slats nailed to them, six packages stamped DEEP FRAMES, three Stop & Shop bags, and six brown paper parcels stamped DURAGILT, the smoker, the hat, the gloves, the hive tool, and the beesuit in its package. The air became rich with the smell of beeswax, one of nature's perfect smells, made, as it is, inside the very young worker bees, from an amalgam of around three hundred ingredients, which end up smelling like a mix of two seasons: autumn, as you walk over dry red leaves through the woods, and summer in a field of wildflowers. The packages emitting this intoxicating aroma were the ones containing the thin yellow wax sheets that would need to be fitted into the frames, the ones labeled DURAGILT.

My bed was positioned so that when I went to sleep at night and opened my eyes in the morning, the first thing I saw was the word DURAGILT, and everything waiting to be assembled. Herein lay the problem. It had to be assembled with one tool in particular, one that scared me beyond reasoned argument. A hammer. I knew I owned such an item for I had seen it in my tool kit when searching for cuphooks, which screw in easily by themselves. I had never used one. Anything that required a hammer was something I called somebody to do. I had never had the slightest ambition to be known as "handy." I used to joke that I would need to remarry when the lightbulbs burned out. A hammer looked at home in a toolbox next to screwdrivers. But a hammer is heavy and metal, looks made for committing crimes of all sorts, and, lacks discernible charm. Nonetheless, this bee project of mine required me to use a hammer to just, simply, get ready to begin. It was a hammer or nothing. So it was nothing.

By the second week of April, I knew it could not be nothing for very much longer. May was on its way. I worked up the courage to ask my daughter's best friend's father for help. He is not one of the local would-be philanderers single moms in small towns tend to meet. He is the kind of guy who leads the twenty giggling little girls of the local Brownie troop, paints like an Old Master, and designs realistically grotesque monster-villains for video games. He didn't like to buy what he could make himself and he could make anything.

He tore open one of the packages labeled DEEP SUPER. The four sides of the hive body were notched so that they fit together and made the top-less-bottomless box from which ten wooden DURAGILT-filled frames would dangle, and in which my future bees would set up house. The boards came with a bag containing the exact number of nails needed. He fit these pieces one to another and bid me open another wrapper and do the same. Then he began pounding and the nails went straight in as I held the sides. He drove five nails in what seemed like five seconds and his rhythm was so steady and it looked so easy that I was secretly hoping he'd just bang in the next thirty-five and the next eighty after that and I'd have three hives built. It would just be a matter of another hundred and fifty nails for the hives and eight nails per frame times ten frames per deep super, with two deep supers for each of the three hives making 480 nails for the frames—not including the five ⁹⁄₁₆ of an inch-long staples to be shot into each of the frames by a staple gun, which perhaps he would also be willing to do because I had never even contemplated seeing a staple gun within my walls, much less aiming and shooting one—and the whole thing would be done and I would applaud and never have to touch that fearsome hammer. But he put the hammer in my hand. He intended to teach me.

The nails that glided for him did not drive so easy for me. I had to swing it steadily. Hit it straight. Use a little force. Not too little. Not too much. Just get the feel. And to my amazement, I got the feel and the hammer played scales as the nail vibrated and pushed into the wood. There was a music to it, especially when the rhythm was right. And I was proud. Until I looked inside and saw that I'd driven it crooked after all and the point of the nail stuck out the other side of the wood like a nasty threat. He took the nail out. He did it in one movement with the slotted thing on the back of the hammer. Then he gave the nail a few taps and made it straight again. I drove it right after a few tries until I drove it wrong again. The frames were harder. The nails were small and skinny and bendy, especially for me. He was kind. I might be having a problem with knots, he said. If there were knots in the wood, anyone's nail might buck. But his nails didn't buck. Knots or not, the frames had to be assembled properly because once they've got the wax in, the DURAGILT, that's where the bees do almost everything that keeps them alive. I, the fledgling hammer-wielder, wasn't going to let any knots stop me now. He showed me how to use that slotted thing to deal with straying nails. I tried to do it precisely the way he did, but those skinny wiggly knot-shy misogynist nails taunted me with their waywardness. So he taught me the wonders of bent-nose pliers and I found a new love. Those curved serrated jaws could gnaw me out of any mishap. The pointy tips nestled under the nailhead and if I squeezed and twisted, I could get any nail out of whatever bad place I'd put it. What's more, there was something artful about the result. The nail took on a new appearance, sometimes wavy, sometimes a spiral, sometimes a simple J shape that would have made a fine fish hook. There was no hope of making these nails straight again, so I took a bunch of good ones to my buddy at the True Value hardware store and had him sell me a bag full of extras. I gave up minding that straightening was beyond me and grew fond of my bent nails. I'd put them in. I'd taken them out. I got a Ziploc baggie and set my mangled mementos aside to keep.

I hammered in my bedroom when it rained. I hammered in the sunshine when it didn't. I hammered for hours and hours at a time and I became a hammer aficionado. One heavy hammer was not enough. The skinny, wiggly nails liked an upholstery hammer and the inch-and-a-halfers liked the midsized compact hammer that had sung its siren song from the aisles of True Value. So I now had three hammers and I hammered with all of them. When I still had trouble I telephoned the Bee Master. His advice was simple: "Buy a drill." The drill would make little holes so the nails would know where to go. I bought the drill. I am not fond of things that roar and have to be plugged in. Not even vacuum cleaners. The thought of calmly and effectively using a drill that not only roared and had to be plugged in but also wielded terrifyingly sharp spark-

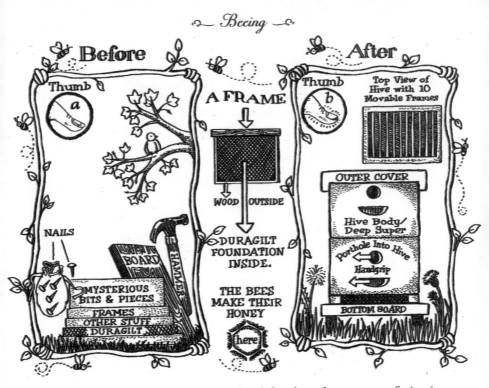

Before

Thumb *a*

A FRAME

WOOD | OUTSIDE

DURAGILT FOUNDATION INSIDE.

THE BEES MAKE THEIR HONEY

here

NAILS

MYSTERIOUS BITS & PIECES
FRAMES
OTHER STUFF
DURAGILT

HAMMER

After

Thumb *b*

Top View of Hive with 10 Movable Frames

OUTER COVER

Hive Body/ Deep Super

Porthole Into Hive

Handgrip

BOTTOM BOARD

spitting metal bits created by mankind for the sole purpose of piercing things was beyond my imagining.

The drill was a talisman. The simple owning of this magic object improved my hammering. I did not have to open the box. It now took me only an hour to build a frame, and I even stapled the DURAGILT so it fit tidily inside the frame top to bottom and side to side and didn't flap about like an embossed wax windowshade. Like the genius I now knew myself to be, I let life's requirements languish as I mastered the pounding art.

Then I slammed my thumb. With the big hammer. The one I'd been afraid of in the first place. My thumb turned cartoon-huge and grape-purple and throbbed up and down to the beat of my heart. Trimmed with yellow and red at the edges, it was the most magnificent trophy a hammerer could hope for, something to wave in front of your friends, a true emblem of initiation into the league. But losing the most valuable fifth of my right hand put me half out of the primate class and into the realm of creatures who must get about their business without an opposing thumb. My spirits plummeted.

No matter how I struggled, my daughter watched me with admiration. "You're doing a good job, Mommy," she said.

"No, I'm doing an awful job." I sighed. "These nails are bending like snakes and they keep coming through the wood and I don't know what I'm doing and my finger hurts and I'm frustrated."

"But you're still doing a good job."

"Thanks, honey, but I'm not. I stink at this. But that's OK," I added, as much for her sake as my own. "I'll learn."

"But you are!" she insisted, raising her voice.

Two days' worth of dishes were piled in the sink. The laundry lay unwashed in five color-coded piles on the floor. My thumb had turned a dull gray-green. There were splinters in my bed, nails on my floor, and my three hives' worth of bees, all thirty-six thousand of them, were just about ready to head north to live in hives that did not yet exist. My sight was blurred by captive tears. I stopped my hammering and challenged my child. "Name one thing I'm doing a good job at."

"Following your dream?"

My daughter's remark was a shock to me. I had, after all, always encouraged her to follow her dreams. Was that what I was doing? Lost in cussing and hammering I had failed to acknowledge that this dream in my house was a dream at all. It was careless, hardly invited, impulsive. But my daughter knew what it was. Now I followed it more and faster. We were three days away from bee day.

3

Bee Days

Anyone starting beekeeping
should begin with three hives,
the sacred number.

—*an old Central European belief*

NYTHING THAT WAS NOT HAMMERING was not done. The sun was warm so I worked outside with the sound of hammer on nail echoing in the valley, as if an invisible companion worked at my side. A few houses away, one woman's husband, brown and shirtless, mowed the lawn over and over again. He confessed to me that it kept him out of the house and away from the kids, which was where he wanted to be. Accompanied by the roaring machine, he found a kind of peace that was his nowhere else. But he was not left to it. The other women on the street hummed about his tanned bare chest and how their pale, round-bellied husbands found his shirtlessness an affront, an outrage, indecent. The bare-chested renegade watched me work. "You planning to keep those bees in the backyard?" he wondered.

"Why?" I teased. "You afraid?"

"I'm not afraid of nothin'." I smiled at the convenience of the double negative and assured him he was safe.

One day away from bee day, an old fellow from town pulled up on his black Harley. It was Herc. Herc has a polished bald head and the muscled body of a man half his age, which was, he told me, "ninety-nine upside down." I'd seen Herc on Main Street. He always wore his shirts with the sleeves ripped out and his strong arms exposed. My daughter wondered where all those sleeves had gone but it was one of those questions you never asked and couldn't answer. He was around a lot because death had stolen his wife five years before and sometimes he liked to talk to someone other than his dogs. We'd exchanged cordialities, no more. But just because I'd never said a word to him about my doings didn't mean he wasn't well aware of them. He swung his leg over the seat of his Harley as if he were dismounting a horse and loped across the lawn. From the way he walked, I knew he had a story. He looked at my work and said, "Used to keep bees. 'Til my wife died. Then I didn't. Lemme do that for ya." He did a half-day's work in ten minutes and another half in another ten, and owing to him the job was nearly done. Wanting to thank him, I had him up to the kitchen and brewed a pot of strong coffee. I cut us both a piece of apple pie. He ate firsts and seconds and thirds and told me his life. His manner was gentle and courtly. He'd been in the army. He'd been a boxer. He'd been to Cuba. And both he and his only daughter were championship ax-throwers, best in the state or damn close.

Rain fell all night. This was good for spring but not for me. If rain is falling on the day you are supposed to set up three hives at the far end of a field, that is not wonderful. If on that same day you intend to install three queens and bees numbering twelve times the population of your town, you have a fine mess. The job cannot be done. You must then put your bees in a cool dry place and placate them with a light spray of sugar water until the weather improves. The only cool dry place to hand was our second-floor flat. They would have to be houseguests. My daughter couldn't wait!

We awoke to rivers in the street and breakfasted in the local diner. My daughter regaled the local Episcopal minister, a brilliant anglophile, with a detailed scheme involving the continuing downpour, plumbing, windows and secret tubes full of bees all of which would be made to work together toward a most satisfying end: driving the skittery landlady mad and scaring the hell out of the landlady's three loud and quarrelsome daughters. I silently wondered if Episcopal ministers had any pull with the weather.

Ever since my daughter weighed less than fifty pounds, she has been drawn to that which grown-ups consider mythology. To her, the mythical seemed more comprehensible than the workaday world. Once she spotted a unicorn behind the Congregational Church. For some strange grown-up reason I could not comprehend, a lot of adults went to extraordinary trouble to convince her that she had not seen what she saw. She stuck to her story. Later she wrote to her unicorn, just to stay in touch. "I still bleve in you," she promised. The antics of Zeus, Hera, their parents and many children entered our lives when she was four and stayed as honored guests. So it was quite logical to both of us that I called on the good will of Helios, the sun, as we drove to the field.

Bathed by a constant drizzle, I carried the heavy cement blocks, two at a time for balance, through the matted yellow grass and the muck to my chosen spot: facing east, near but not under trees, next to abundant fresh running water. The only way it could have been farther from the road was if I'd sited it as the ancient Egyptians used to do, on a barge in the middle of the river. Still, I was innocent and happy and the drizzle seemed a touch lighter than before. I hauled the hive stands next. Since the inside of a bee's home must be dry, it didn't seem the moment to set up the hives. Returning home, I fitted ten wax-filled frames into each hive body, gathered the tops and bottoms and laid out my hive tool, my smoker, and my unsuitable hat. At last, I opened the package that held my beesuit. A piece of paper fluttered from the pocket. CAUTION, it said. DISCLAIMER OF ALL WARRANTIES. Discarding any confidence I might have had, I read on. The manufacturers of the beesuit wished to advise me that though this heavy white cotton garment might protect me from bee stings, it also might not, and not to come crying to them if it didn't work. They wanted me to know that I, "the purchaser," had purchased these coveralls "without any implied warranty of fitness for a particular purpose." That was reassuring. If something isn't fit for the particular purpose it's made for, for what is it fit? I was on my own, unpredictably unsafe.

It was time. I set off alone; abstractly excited, absolutely afraid, yet suddenly cloaked by an astonishing calm. Tranquility was the last thing I'd expected. My car became the vast quiet place between now and the next thing. Fear could stay as long as it didn't make a fuss. Twenty-three miles passed in a half hour, a minute, a day, a year. I had no sense of traveling at all, but I'd arrived. A van, three trucks, and a car or two were crammed into the small parking lot in front of the beekeeping supplies store. I found myself thinking that if all these people were here, then I must be here too which meant that, No, I was not imagining this whole bee business, an idea my independently unconvinced mind apparently still nourished.

At right angles to the little white storefront was a wheeled flatbed tented in canvas. It looked like an old-fashioned circus car with the paint worn off. The canvas had been drawn away and tied back, to the delight of those gathered. The rain gave pause. Recalcitrant Helios came out from behind the clouds to take a peek. The bees from Georgia had arrived, full of life and singing out loud. There were hundreds of thousands, maybe millions of bees, apportioned off in neatly piled wooden and wire boxes. Along with ten to twelve thousand subjects humming like a chorus finding a note, each box had a queen safe behind a marshmallow candy door, tucked into her own small throne room and ready for her new kingdom. In the midst of this, the Bee Master stood, an ebullient ringmaster dispensing the same good advice over and over as he cheerfully wielded huge, red-handled clippers. I floated through the next few minutes as if I had happened upon a mesmerizing sacred dance to which all the natives knew the steps. When it came my turn at the front of the line, the Bee Master counted three boxes and briskly snapped the wood that bound my bees to the others. Trying to disguise my own awed unbelonging, I smiled, but I kept my lips closed for I noticed that along with the bees buzzing around a deliberately leaky tin of sugar syrup in their tidy cage came an accompanying entourage of uncaged bees who were eager to explore the world around them. I did not wish to swallow one. My bare hands clasped the wooden bars that both separated and joined the three future colonies. My burden was unexpectedly heavy. I held them two-handed in front of my knees as I tried to absorb the Bee Master's rat-a-tat primer on what to do with these creatures. OK. You sprayed this, you banged that, you pried this, you pulled that, you stuck something here and twisted something there but not until you removed something else and banged again. Then you dumped the bees in and raked them with your naked fingers extended.

I hobbled to my blue Honda. "I've never done this," I bleated to a bespectacled bearlike man with hair the color of Hungarian paprika. He

was holding his own buzzing wooden box easily, as if it contained choco-
lates. "I've never even seen a beehive close up."

"Then follow me," he replied. So I did the only sensible thing. I settled my bees in the backseat without benefit of seat belt, opened one of
my windows a crack in cast anyone wanted to leave, and followed a large
man I'd never seen before to a place I'd never been before while pretending that the curiosity of a couple dozen entourage bees, which caused
them to freely explore their coach and driver, was perfectly fine with me.
When we arrived at the site where his new hive was to be, which was,
by no coincidence, behind the house that served as his office building, he
introduced himself. His name was Red. "Put this on," Red commanded,
as he tossed me his grizzly-sized beesuit and helmet. I put my nose up
close to the hive while he spritzed and whacked and poured. Then he
said, "Now just go do the same thing three times." He squinted at the
darkening sky. "Hurry!" He handed me his sugar-water spritzer and his
business card.

I ventured one question. "How do I get these boxes apart?"

He replied in song. "With a saw, dear Liza."

A saw. I thanked him and waved and promised to be in touch and
smiled a closed-mouth smile as the bees measured my curls. I didn't own
a saw. It took me a half hour, speeding, to get to the True Value. I dashed
in. "I need a saw fast." My True Value buddy doesn't go fast. "You
buildin' somethin'?" I led him outside. A small edgy crowd had gathered
round my Honda. "You get bit yet?" he wondered aloud. I was watching the pewter clouds close in on the valley. He watched the bees wander across the upholstery. "Better mind where you sit," he said, drawing
a laugh from the onlookers.

Finally, he moseyed back into the store and returned with something called a junior hacksaw. Eleven dollars and seventy-nine cents later, no obstacle remained but myself and the roiling heavens. I hurried home, loaded my car, and put on The Outfit.

Back and forth I shuttled across the field, a mound of black clouds gathering directly overhead, setting up the boxes and putting in frames. Ignoring the sticky puddle of sugar water in my car, I pulled the wrapping off my saw. I slid on my ghastly hat and zipped myself into the netting and went at the bars connecting the boxes with judicious vigor. Done! I pocketed my hive tool, some garden wire, and three blue thumb-tacks. Two more trips and the bees were at the hives. I felt a drop of rain. It's just one drop, I told myself, and gave the bees another squirt of sugar. Then I whacked the box against the ground and the bees dropped to the bottom just as they were supposed to do. I pried the wooden door off the first package. The queen cage was packed next to the feed, a heavy, drippy, sticky supply of yet more sugar water in a one-pound coffee can with more holes than the SMC Cartage Company garage after the St. Valentine's Day Massacre. I struggled to remove it, just catching the queen cage before it dropped unceremoniously into the masses, and slammed the wooden door back over the top. I removed five frames from the hive body and rested them very carefully on their sides against the hive stand. Then I pulled the door off a second time and turned the whole cage upside down, shaking thousands of sugar-sated bees into their new home. Not everyone was willing to leave, so I put the cage in front of the hive with the hole facing the entrance. Now, timid or not, I slid on my gloves. I reached one hand into this humming black pile of stingers and wings and gently spread them out across the bottom so that they would not be squished to death when I replaced the frames, which I had to do next. I allowed myself a moment to breathe, but nothing more. I pressed a blue tack into the fifth frame and one into the queen's cage. Using the garden wire as a holding rope between the tacks, I suspended Her Majesty between the fourth and fifth frame so that she would be safe while her new subjects came to know her.

Rarely is a queen the mother of the bees that come in a purchased package. If she were rudely dumped in with her new subjects, they'd more than likely kill her. Hence the marshmallow gate. As she and her attendants eat their way through, her subjects nibble the other side. In the few days it takes to gnaw her way into the hive, her pheromones, her unique message-sending scent, will become the pheromones they honor and heed absolutely and she will become the perfectly transplanted heart of this new body.

As I did this work, quickly, but without hasty movement, without frenzy, gliding through the ritual a second time and then again, I experienced an awe that is as old as humanity.

The moment I had feared was over too soon. I wanted to go on and on being with these strangers but the rain began to soak my beesuit. I lodged up-ended jars of sugar syrup I'd prepared at home into feeders that were themselves lodged into the fronts of the hives. The sugar syrup, a one-to-one mixture of sugar and warm water I'd prepared in a lobster pot on my stove, is a gift to the bees so that they have ready nourishment and can devote their energies to turning the wax foundations into useable honeycomb and begin rearing brood. If they had to start fresh and be responsible for finding all their food, it would slow things down mightily. I moved the cages with stragglers closer to the hives and turned them wood-side up hoping the slow ones wouldn't be soaked. They had to travel this last leg of their journey from Georgia on their own. The rain began to drum on the galvanized aluminum hive covers. Leaving the hives, I crossed the field slowly, walking half-backwards, exultant and sad, stunned but unstung. I unzipped my netting, removed my helmet and stood, letting the rain pelt my cheeks for as long as I could.

A very short time ago, I would have said there was as much chance of my keeping bees as of my climbing Everest. With a minimum amount of gear and absolutely no qualifications, I had undertaken it. I gladly humbled myself before the greatness of nature. Like Sir Edmund Hillary, I had become a beekeeper.

That night, I was alone with our old man cat, Ruffy. I sat as still as he did, thinking only of my bees, and listening to the thunder as if it were a strange sound I had never heard before.

How must it be for such tiny creatures, dumped into a new home in a strange place, unsure of their new and foreign queen? How must it be if you are confined by night and weather inside a kettledrum, tormented for hours by the tympanic rumble of hard rain on a metal roof? And what of the stragglers, the ones I'd left to find their way through the storm? Was it too much to hope for their survival? Sleep was impossible. Waiting for the dawn, I hopefully prepared my "baby bottles." Using the old black enameled lobster pot, I mixed five pounds of sugar with five pounds of water—one pint equals one pound—over low heat. When the sugar dissolved into the water, the milky liquid turned clear and I shut off the burner. No need to test this formula with my pinky. The bees take it

room temperature, from the hole-peppered lids of upside-down Ball jars. I ladled the syrup into jars. If I could not assure them gentle weather, this, at least, I could do for my poor little charges.

At dawn the only sign of the violent night was the darkened pavement of my driveway. Nature, reckless and carefree, had moved on.

Bearing Ball jars filled with sugar syrup, my daughter and I crossed the field. The gray-greasy-green mud sucked at our colorful Wellington boots, pulling against our eager stride. The hives still stood where I'd put them. That, in itself, was a wonder to me. The wooden boxes were empty. The strays had found their way. All by themselves. What exceptionally clever bees we had. My daughter thought we ought to name them. She proposed the queens be called Queenie, Queenie, and Queenie. That part was easy. The thirty-six thousand other monikers were hard.

Crouching beside these three mysterious boxes, we agreed these little principalities ought to be called something other than this hive and that. Something important. I had decided upon three hives because the Bee Master said that three was a good number and he trusted it. If one died out, I had two. If two, one. A simple upping the odds. But three is not your run-of-the-mill integer. It holds an auspicious place amongst the primary numbers. Pythagoras thought it perfect, a symbol of deity in which might be found the beginning, middle, end. Our three were all beginnings. Who knew whether they would be three Furies, three Graces, three Harpies? Three stood for not only the holy Christian trinity, but another, ancient and divine: Zeus, Hades, and Poseidon. Red, yellow, and blue.

Body, mind, and spirit. Faith, hope, and charity. Animal, vegetable, mineral. World, flesh, and the devil. And the three ships that carried Columbus and his men to the new world, the *Nina,* the *Pinta,* and the *Santa Maria.* Later, rummaging through a dictionary of mythology and fable, we came across the "Mysterious Three" who guard Valhalla, the Viking heaven— Har, Jafenhar, and Thridi—meaning, in order, High, Equally High, and The Third. The hive nearest to the river was christened Har. The middle hive would be Jafenhar, and the one nearest to the road, which was still a long way off, would be Thridi, the third.

Having lingered until our eyes blurred, we ventured south to Stop & Shop to buy more sugar. The bees had to cover those DURAGILT wax foundations I had struggled over with perfect little six-sided tubes called cells. Business in the hive couldn't really thrive until that was well begun, for it is in the cells that the queen lays her eggs, the brood is reared, the pollen stored, and the honey made. I figured if we bought the groceries and did a bit of the cooking, the work would go faster. As we loaded the blue Honda with forty pounds of white refined sugar, a little bee flew out of the car. This adventurous spirit had come for the ride. She lighted on the front door as we loaded, and when it was time to go, I ushered her back inside, not wanting to leave her fifteen miles from her new home. She rode on the sunny platform behind the backseat and we dropped her off at Farmer Tom's field, where there was no sign of Farmer Tom.

The next day we saw something we had not seen the day before. A honeybee has six little legs, which, at the start, look pretty much the same, thin and blackish with what look like two tiny joints. Looking closer, we began to realize that they were not really all the same, but graduated in size, with the back legs looking almost sturdy. The bees that left the hives looked the same as the day before, but many of the bees returning from their travels were dusty-looking and bore butter-yellow balls stored tidily on these back legs. This was pollen, and they'd gathered it to themselves, filling their tiny pollen baskets for the good of the hive and umpteen flowering plants that generously bestow both food and beauty upon the world of man.

In this part of the country, at least 80 percent of what was once the wild bee population has died out. What these bees were doing was not only good for their own hives and the local crops, it was good for the lovely flower and vegetable gardens of the anxious wives who had been good enough to ostracize me. We, my daughter, my bees, and I, were doing our chary neighbors a kindness they would never recognize, but a kindness nonetheless. The bees were doing urgent work. Whether or not the human world had a high opinion of them or, in their ignorance, thought they were best restricted to metaphors, needlepoint throw pillows, and cutesy coffee mugs, the work would go undone without them. There would be no strawberries. There would be no cucumbers. None of the hundreds of fruits and vegetables we would grieve to be without. Not that the bees knew or cared. They lived fully within their own world, parallel to mine but to which I was almost incidental. Yet, in my world, they were deeply important to me as well as those hostile toward, or mindless of, their existence. Their gathering was underway, though the earth more suggested spring than declared it. Up to the moment I experienced the particular thrill of seeing my bees doing what bees do, I thought of pollen as that which gives you an evil headache on a sunny day. Now I had an investment in the stuff and it almost seemed as wonderful as it truly is.

I don't think of myself as patient, but I had to be. Four days would need to pass before we could peek inside, check on the drawing out of the comb, and pay our respect to the queens. Any sooner might mean an uprising.

In addition to the magic curtain, my kitchen has two windows. One looks onto the space between our house and the one next door. The other looks across the street at the gingerbread cottage of a pistol-packing sociopath who worships his chain saw and hates kids, and the compact colonial that houses searingly witty Mary and her tribe of eight children and a husband who dyes his hair. Everyone knows that he dyes his hair because he takes no care to match the last shade with the next as long as it covers the gray. He is a handsome man with lust in more than his heart. Though Mary, who is built like a fire safe with great legs, seems to turn a blind eye to her husband's wandering one, I have heard her declare that she aspires to be the town gossip. She can't stand to see a fire engine without knowing what's aflame. She needs to know who is getting mangled in what car crash and who is brawling at the local bars with such a passion that one Christmas she asked for and got a police scanner so she could sit in her house minding other peoples' business. When I chanced to look up from the kitchen table and out the window and saw the lace curtains on Mary's front door discreetly pull aside for a peek, I knew something was up. Moments later, there was a knock at my door.

It was Herc. He was wearing a plaid shirt with the sleeves ripped off to show his muscles. He was bearing an armload of lilacs for me and a chair he'd cut out of a log for my daughter, a small chair with a teddy bear strapped to it. Astonished, I showed him into the house, whereupon he stammered very sweetly through a well-thought-out declaration of love.

"I knew I made good pie, but I had no idea it was that good," I answered, for want of anything else to say. I have always thought that "flummoxed" was a fine word—with that "ox" in the middle—for which I had no use. Until that very moment. Flummoxed was what I had just become. I cannot say it is ever pleasant to dash a man's hopes. All the worse if it is a lonely man with an aching heart. But if that man is also a hugely muscled championship ax-thrower of unknown temperament, there is a certain sense of alarm. I scurried busily around the kitchen fussing with coffee cups and beans and water and muttering something about him being very sweet. The splendid perfume of lilacs filled the kitchen and we paused for a moment to breathe it in. I poured coffee, he sipped. My mind frantically tried to compose charming, unhurtful refusals, hampered by the persistent image of myself from the back, pinned to a tree by an ax still vibrating from the force with which it had been hurled. Nothing else came to mind. Nothing is what I said. He asked for a second cup of coffee. My curls were matted, blood drizzled between my shoulder blades. Compared to the straightforward business of dipping my hand into stingers and wings, this was the more strenuous occasion.

The next morning, it was time to check on the queens. I suited up in brilliant white. I readied my smoker, placing twine, twigs, and old pages of the *New York Times* in a lavender Bergdorf's shopping bag that was the perfect size. I set aside my canvas-topped yellow rubber gloves and a pair of wrist-length flowered kidskin gardening gloves. No matter that a casual glance at Charles Butler's *The Feminine Monarchie,* written in 1623, warned against "leather in gloves" for "as soone as they touch it they will strike, if they be any whit moved," and no matter that the most current books I had to hand said much the same though not as prettily, I decided that my particular bees might find the flower print a cheering thing and I wanted them as happy as possible. I packed a dental pick I'd purchased at the True Value because it looked like a good item to have at the ready for what reason I knew not, and I brought my sugar-water cocktail.

I had studied up on the smoker. It was supposed to go like this: a layer of crumpled paper at the bottom, then some broken-up twigs to move the flames on up, twine, to burn slowly and make a lot of smoke and then, the match. The bellows attached would bring air to the mix; and once the fire was crisply underway, you garnished the top with a handful of grass, meant to cool the smoke so that the heat of it did not scorch the bees.

The purpose of a smoker is to cry "Fire!" in a crowded hive. A few well-placed puffs and the bees are in crisis. They hurry to save their labors, engorging themselves on honey in preparation for a mass evacuation. In this state of dismayed activity, they are naturally less likely to pay as much attention to defending their hive against a beekeeper. It struck me as an unkind thing to do, but I hadn't a better idea, so I hunched over at the edge of the field and struck a match. Then a second. A third. The air was gluey hot and moisture was rising from the ground. There was no wind. I struck a fourth match and the paper caught. I pumped the bellows but the flame expired. I repacked and struck thirteen more matches until there were no more to strike, at which point I decided to take my chances.

I zipped myself in under my dreadful hat, gave the bees their sugar-water breakfast, and held my breath as I lifted the cover off Har. I had been told that when working bees, all worldly cares dropped away. Now I knew why. When you dip your hands unbidden, with or without flowered kidskin gloves, into a mass of bees, you had best honor the present with your total attention. Focusing beyond the irritated insects ringing my arms and face, I quickly unwound the garden wire that held the queen cage to the tack. I held it up to eye level. The queen had been released but she did not present herself and thus I could not perform my deep curtsy. I could only hope she was alive within and had begun her reign. I slowly pressed together the frames that had lodged her cage so that they fit as they are supposed to. I moved slowly, so that the bees who had wandered between the ribs would sense the diminishing space and go elsewhere before they were squished. I knew that one way or another it was likely bees would die every time I invaded the hive, but I did not want this to be true.

I do not know how long it took to do this small job that felt very large to me. I cannot imagine it was very much more than five minutes. But I could be wrong. My concentration was so complete that time took a less recognizable shape, as it does in the vast cosmos. Time passed through my fingers and my eyes, my concentration made it dense and uncountable. The bees tolerated me with great patience. It was the same with Thridi, but with Jafenhar, the middle hive, I had a start. The blue tack dropped between the frames and vanished silently into the humming dark as if down a bottomless well. With the tack went my magical calm, but only for an instant as a young bee explored the gap between my glove and sleeve with her stinger half-extended, tickling my wrist unbearably. I waited for four-hundred-years'-worth of wisdom on the subject of leather gloves to prove itself correct and I was pleased that it did not. The tiny wire-faced boxes with three round rooms, each the size of a nickel, were my trophies and I bore them home proudly, setting them on the dinner table to be admired.

Of all the sorrows the Olympians endured, none touched my daughter as profoundly as the loss suffered by Demeter, whose only daughter Persephone had been kidnapped. The pain of separation between mother and child was something she could well understand. For what if she lost me? What then? I tell her that I will not leave her. That if some day I am taken, it will be only by death, and then I swear that my love and spirit will always be with her. In the meantime, she will come to see that she has herself to depend on as well. But when you cannot reach the glasses in the cupboard without a chair, that idea is hard to snuggle up with. I feel for Demeter myself. Spring was when Demeter's wintery mourning came to an end with a mother's embrace as Persephone returned from the underworld.

All of a sudden, grass seemed to grow in a day. A hawk trawled the great gray sky above the river. My windshield was covered with a fine yellow dust. Pollen fell from trees whose leaves yet only suggested themselves. From early morning, tractors roared through the fields around the edges of town. A local farmer died. His tractor roared behind his hearse, ahead of his widow, as the procession made its way down Main Street.

Even in death, there was homage to rebirth. Almost everyone readied the soil in their gardens for planting. Each day, I waited for mid-morning when it was time to visit the bees. It seemed to me the bees had come to expect the prompt arrival of their caterer. When I would step out of my blue Honda bearing the sugar water, I would inevitably be greeted by one envoy, traveling alone. She would escort me across the field and remain at my side as I removed the empty Ball jar, carried it away from the hives, and prepared the new one. And then she would escort me back to her hive or hover while I tended to the others. Was it the same bee every morning or was it simply the nearest bee sensing promise in the form of this oversized multistemmed flower that regularly arrived in a moving pod and lumbered across the field laden with nectar? I cannot say.

But it was always just one, and it was during these first timid comings and goings that I began to recognize the obvious. Up until then, I'd thought of bees the way some Americans tend to think of the Russians or the Chinese, as a large quantity, an edgeless blur of foreign Others. Now I saw individuals. To even begin to really see a small bug in motion requires several lookings, and one of the first things I noticed was this: Neither the bees that inspired all the bee chat that prompted me to purchase antique volumes by Maeterlinck and Fabre nor the bee that had anchored me in the present and reminded me to "just be" during hard times had anything to do with honeybees or beekeeping other than that they, along with more than two hundred thousand other species of winged insects, belong to the order *Hymenoptera*. The bees that had flustered the wedding guest wearing the bee brooch were actually wasps, yellow jackets, and the photograph on my treasured postcard was of a weathered bumblebee well past its prime. I had been misinspired. Now it was too late. I could very nearly hear the gods laughing at me. I some-times think I live to entertain them.

As they laughed, I watched. I saw how often one bee landed face to face with another and how they conversed, feeling each other quite purposefully with their antennae. I saw one bee in Jafenhar struggle to extract a dead sister from a small space between the feeder and the hive body. She pulled and pulled and pushed and nudged and her thin black arms were still at the task minutes later. Her struggle touched me and I tried to help. Meaning to unblock her passageway, I pushed the carcass out for her and it fell onto the bottom board, but that didn't settle the matter. For whatever reason, the dead bee was then pulled back into the hive. Another bee dove into a jar of sugar water before I had a chance to cap it. She paddled frantically, trying to right herself. Her life clearly mattered to her and she fought for it. I offered her my hive tool and she climbed aboard. I set it down and continued my work, but when I returned she was still there. I wondered if her wings had become too heavy from the sticky syrup. I brought her to a purple wildflower. After several tries, she attached herself and I hoped for the best. She remained on the flower for a short while, then fell to the grass.

I felt a maternal ache. Scientific detachment was not to be mine, nor did I wish it so. I was a walking, talking demonstration of Heisenberg's uncertainty principle: My very presence as an observer of the bees, not to mention my placing them in hives so they could live and make honey, was meddling that altered everything I would or could observe about them. Whatever I saw was affected by my presence in some way, even if I kept my well-intentioned hands to myself. And so what's a mother to do?

I watched one pollen-laden worker confidently zoom toward her landing strip with her tangerine cargo only to totally miss the landing and do a backwards ally-oop onto the grass. She squiggled herself aright and launched herself successfully the second time around and marched into the hive with her booty. Satisfied for the both of us, I went home to tend to my human young.

In my town, I'd wager, not too many local farmers have any reason to think about the profound mystery that spring once was. How was it that grain, dead all the winter, then sprung reliably to life and fruitfulness each year? The Greeks told the story of a mother's love for her daughter and how, only upon her return from unsought darkness, once and every year thereafter, would the world rejoice with assurance of food and fecundity. Those that farm up this way are more connected to the ancients than they might ever suppose: Custom dictates that there be no planting before Mother's Day. No matter that there be three weeks of warm, sunny days beforehand. They wait. After Mother's Day, there is no more fear of frost. Demeter gets her due; latter-day mothers get their store-bought roses and breakfast in bed.

Finally I happened to see Farmer Tom at work in the field. By his side was the blond-haired lady who was his companion of fifteen years. Longer than I'd been married. I wondered why she was not his wife. They had roped off a garden-sized plot in the midst of eleven acres allotted for the farm and linked up a long chain of hoses with a lawn sprinkler at the end. Within the yellow-cord boundaries they had cleared boot-width walking paths and dug little furrows. I walked over to them and saw that behind the barns, Farmer Tom had set up several boards on sawhorses. Upon them, he told me, were seven thousand dollars' worth of seedlings, an entire farm on a tabletop. Farmer Tom did not seem particularly prepared. He owned absolutely no modern farm equipment, not even a rusty truck. Kneeling beside their small furrows, their faces covered by sunhats, Farmer Tom and his lady dug with rusted tools and patted young broccoli rabe into the soil as yellow butterflies rimmed with black danced above them, mapleseed helicopters spiraled madly, and the river ran a stern military green.

In my kitchen, I measured out portions of my new life with pounds of sugar and Clorox. Everyday I noted how much sugar syrup had been taken by my girls—already I had come to call the bees my girls—in Har, Jafenhar, and Thridi. I tried to figure out why one day they ate heartily and another they picked as if I'd fed them vinegar. As soon as I had crafted a well-reasoned theory, the next day's evidence would prove me completely wrong and I came to delight in the girls' inscrutability.

It was this quality that charmed me most of all. In their simplicity, they would not be obvious. They could not be summed up.

In the morning as I packed a school lunch, I melted sugar in water on the stove. In the late afternoon, as I negotiated the sugar consumption of my daughter, I would do my daily hot wash, delighting with unseemly glee in that which I had always sought to avoid—a bleachy smell and shrinkage.

My beesuit was as whitely blinding as glare upon the sea and the crotch was creeping up to the right height with every savage laundering. It was becoming a masterpiece of the laundress's art, but when it came time for my long anticipated lesson with the Bee Master, I knew enough to leave it home or face searing, if silent, ridicule.

The wearing of a chic white uniform is, though nobody says so outright, perceived by some to be rather sissified. It smacks of protecting yourself, which smacks of fear, which smacks of unmanliness. So it's best to appear a bit careless. Long sleeves will pass. So will tucking your trousers into your boots. The plastic pith helmet with mask is fine, but gloves spell coward. It is said you can handle things better with your bare hands, which is probably true.

I spent two minutes getting into my khakis and tan shirt and at least ten more achieving the required barehandedness. On my right hand, I have worn a heart-shaped diamond ring every day since it was given to me by my stepfather when I turned sixteen, excepting the years of my marriage. It has made itself completely at home on my finger. Removing it feels like removing a part of myself. On my honeymoon I slipped on the rain-slick metal staircase of a Parisian *bateau mouche*. I grabbed for the railing and the ring caught and twisted tight. It was either my finger or the sparkling heart. I was hustled into the hold of the boat where the river police snipped off the band with shears. That was the end of the heart for all my married life. When I got divorced, I decided to give myself my heart back and have it fixed. This was to cost ninety-six dollars, which was all the splurge I could afford. On the appointed day, I came to the jeweler to collect my ring. I was handed a bag with a broken diamond. It was cracked, he explained, and it shattered as he worked. I was sick with the symbolism of it all until he handed me the rainbow-flecked heart. Whole. Ring intact, six diamonds around, with a beautiful new diamond in the center. Free of charge. I slid it on and there it stayed until I wiggled and struggled and pulled it off because the Bee Master had made plain the stupidity of introducing bees to girly things like perfume or sparkling jewels. The girly bees find them especially attractive. Not what you want.

The Bee Master acknowledged me with a nod. His smile was genuine, not a smirk, so I knew I'd passed the sissy test. Important. I could tell he had no time for wimps. He dabbed at his sweaty brow. His khakis were rumpled and his shirt was damp. He was wearing his white plastic helmet with the mesh pushed up. He'd come directly to my lesson at his hives from a wasted house call, tracking down a swarm that turned out to be wasps. Dispensing with amenities, he began. "How long does it take a worker to hatch?"

I panicked. "Fifteen days?"

"Don't ask. Know." His tone was firm. He had expectations of me. I had one choice: to meet them. "A queen takes fifteen to seventeen days. A worker?"

"Nineteen, right?"

He would accept that. "Nineteen to twenty-two."

"Why does a worker take longer than a queen when a queen is bigger and has more to do?"

"Think. How many queens does a hive need?" He paused to let me puzzle it out myself. "Only one. But if your bees are making themselves a new queen, they need her in a hurry or the hive dies. A drone takes twenty-four days. What's his job?" He chuckled. "Sex." We climbed a small hill up behind his store and found ourselves at the edge of a field dotted with brown hives.

"People worry too much about manipulating things. When to move what. Rules and regulations. What you need to know is your bees. See things from their point of view. Think the way a bee thinks and then ask, How can I help them?"

To my untrained eyes, his technique for lighting the smoker looked exactly like my technique for lighting the smoker. Only his worked. He puffed cool smoke onto my hands declaring that on such a nice day smoke probably wouldn't be necessary. Reminding me that we work the bees to the back and side of the hive, never blocking the entrance with our bulk and shadow, he opened one of his many hives. With a scrape of his hive tool, he removed the first comb. It was nothing like my empty frames. It was so thick with bees it was hard to say what was underneath. Gently, he reached around to the front of the hive and propped the comb against the leg of the hive stand away from where we were standing. "I've done it other ways," he told me. "Felt those bees climbing northerly along my leg. Led to a few anxious moments." He removed a second frame and showed me how the bees were engorging themselves with honey. "This is what happens when you smoke 'em. Recently I see it when I don't smoke 'em. Why? I haven't seen it in books. But there it is. Maybe there are reasons beside the smoke. Thirty years. They still surprise me."

And then he placed the frame in my bare hands. It was so heavy I nearly dropped it. The bees danced over my naked fingers and I pretended there was nothing unsettling about that. I refused to show fear in front of the Bee Master. My fear, quite present but wholly irrelevant, took itself elsewhere. This was no time to examine myself. I was holding a part of the world I had never visited in my two hands. The inhabitants of this foreign place did their best to continue their work despite my intrusion. No one stood idle amidst the architecture that has titillated the mind of man for at least two thousand years with its geometric elegance. Mathematicians and biologists have tested and tinkered with the honeycomb pattern, neatly formed hexagonal cells tilted at just the right thirteen-degree angle to prevent honey from dripping out. Many have tried to figure out whether there was, perhaps, a more efficient way to do better than the bees, packing more in less space, or, one might say in mathematee, "divid[ing] a surface into regions of equal area with the least total perimeter." On the brink of the new millennium, a mathematician at last "proved," though not necessarily to everyone's satisfaction, that the honeybees knew what they were doing lo, these past twenty-five million years. The honeybees I observed needed no validation from an outside species.

The orange pollen I had lately seen entering my own hives lay packed at the bottom of these perfect cells; in others, I saw butter yellow and the more vivid yellow of a child's crayon. I saw a red the color of merlot, and wondered aloud what that might be from. The Bee Master told me that pollens can be individually identified under a microscope, but that, if I was really curious, I might try sticking my fingers in flowers and seeing what came out. He showed me the differences between capped cells. Honey cells are capped when the honey is set for curing. You can also tell the sex of the yet-to-emerge bee by the cell capping. The workers' cells are gently rounded, the drone cells bulge out a bit. The Bee Master showed me how to turn the frame slowly, as if turning the fragile page of a very old book, so that I could examine both sides without making the bees woozy. I held the frame up as I watched a bee work her way out of her cell with her front legs, her little antennae waving as she was born into light and not darkness. I beamed, awed by what I was witnessing, awed by what I was doing. The Bee Master bent over the hive and deftly lifted a frame from the center. "And there's the queen," he said quietly. She was slim and golden, longer than a drone, delicate, unmistakable as she raced across the combs as if being pursued. But it wasn't pursuit that sped her along, it was purpose. The Bee Master tenderly replaced the frame. "Don't want to squish her," he observed.

It was essential, he added, for me to be able to identify day-old eggs, for it is by the sight of these that you know a queen is well and at her work. He showed me how to stand with sunlight over my shoulder and, when I failed to see what I was looking for, he took the frame from my hand and spotted several. I tried again, looking for what he described as a tiny grain of rice. In solemn earnest, I squinted over the frame, angling this way and that, proving nothing to myself except that I had not been insanely extravagant wasting four hundred dollars in my purchase of swirly blue French reading glasses with a mere magnification of one, but wisely prescient. For instead of sitting on my bedside table waiting for a rare quiet hour when a mother can read to herself, they would now be brought to the field and put to practical apicultural use. I tried again as the Bee Master murmured encouraging words. Impossible. Impossible. And then, perhaps. What I thought I saw was so pale and subtle that beside it, a grain of rice would seem obese. What I thought I saw I may have utterly imagined so as not to appear as stupid and blind as I felt. Nonetheless, I exclaimed, "I see it!" and was promptly stung on the fourth finger of my left hand. No sooner had I gasped, the Bee Master deftly whisked the white pulsing abdomen off my skin with his hive tool, stopping the flow of venom. Speechlessly gaping at my hot finger, I continued to hold the bee-filled frame. I was more stunned by the fact that it hadn't really hurt than by the sting itself. My landmark Genuine

Beekeeper Initiation Sting was nothing next to my purple thumb. It lacked the proper historic feel. But there was one bonus. The mildness of my reaction passed for bravery.

"You're a natural," said the Bee Master, admiring my body's response as if I'd had something to do with it. I basked in his unearned praise.

As I walked my daughter home from school, I announced that I'd received my first official sting. She demanded a look. Our crossing guard, a heavyset, affable jack-of-all-trades called Pumpkin Pete, was standing in earshot. Pumpkin Pete quit this job once every school year looking for something better. But the school always held his place and he always came back. "I heard you got bees," he joined in, and held up his STOP sign while I showed them a painless purplish-red line about an eighth of an inch long and stressed the amazing good fortune of having practically beeproof skin.

"That can change, y'know. Where you keepin' 'em?" asked Pete as he saw us to the far side of the street. I told him. The spidery veins across his cheeks turned rosier than usual. "I wanted to farm that land and they wouldn't let me. Wanted someone organic. I told 'em I was organic. I always been organic. But they gave it to that other guy don't even live around here. Heard he spent himself ten thousand dollars on seedlings." He snorted and shook his head at the wonder of it. "I coulda done up the whole field in pumpkins for twenty-five bucks' wortha seed. But they gave it to that other guy." He sighed and headed back to the middle of the road to shepherd a small flock of children. "The bees woulda been great for them pumpkins," he called over their heads, "and vicey versey."

The next morning was wet, and though I had planned to put my lessons to immediate use before I forgot everything the Bee Master had told me, I could not proceed. Rain makes bees glum and irritable and there is nothing to recommend a visit that dampens hives that need to be dry. It was raining so hard I offered to drive my daughter the two blocks to school, but she was eager to put a new pansy-print umbrella to the test.

Stepping outside, I noticed what looked like a piece of paper folded inside a taped sandwich bag. It rested on the driveway in front of my car. Assuming that one of the landlady's squalling daughters had protected, then dropped, her homework, I picked it up. I crouched so my daughter could hold her pansy umbrella over my head as I opened it. This, spelling intact, is what it said:

"Hi waived to you as you were waling from store Monday. I was driving blue truck Turned around by school to say hello but you were Gone. Wasn't sure what house you livied at. Wood like to Ask You to dinner Friday night at Ram's Gate Inn 8: o'clok Guess we could call this a blid date. Seen you in town often but never had the chance to say Hello. Hope to see you there Al"

"Who's Al?" asked my daughter.

I had no idea. There was a woman called Al who worked the counter at the Central Market, which is where the town gets its cigarettes, newspapers, and lottery tickets. I knew she wasn't the Al in question and I also knew that she was likely to know who was. "I have a secret admirer," I said, and handed her the note.

She read it and nodded knowingly. "He's been asking about you and I told him I didn't want nothin' to do with it. His wife's a drinker and she goes wild if he so much as looks at another woman." I offered my sober thanks and went directly to my favorite neighbor, whose family goes back as far as families go in this part of the country. Had I seen the tough-looking woman who went around in long braids and an Indian headdress with feathers? That was his wife. She called herself She Wolf, the Indian Princess, but everyone knew she was mostly Italian. She'd been arrested a couple of times. Fights. Knives.

What are stingers compared to this? I thought. I was beginning to see that the cliché of the lonely beekeeper might not be what it seemed. Maybe the man or lady in question was merely in search of relatively safe, agreeable company, and knew where to find it.

It rained and rained. Though it is wrong to begrudge farmers rain in the springtime, it was hard to wait for a sunny day when I could actually open my hives. I fussed over their feedings, recording the ounces they drank, and, as if to be ready to seize any moment, I kept my jewelry off. Finally, my naked-handedness paid off. The morning began brilliant and got better. I walked my daughter to school. As the perfect hour approached, an onslaught of anxiety emptied my brain and slowed my motions to the molasses point. I was scared to open the hives. Being brave in front of my daughter or the Bee Master was one thing, being brave in front of myself, another. I wanted nothing so much as to bury myself under my blankets and hide. I cajoled myself. I scolded myself. I had a second and third cup of tea for courage and wished I were the sort who could down a morning whiskey. Finally, I snailed my way out of the house and to the farm and patiently talked myself across the field.

I managed to cultivate a respectable blaze in the smoker and on the strength of that good sign, resolved to remain barehanded even though no one was looking. It seemed natural, being a lefty, to start with Har, which, being on the riverside, was farthest to the left. I smoked the entrance and removed the outer cover. I smoked the oval hole in the inner cover, which I had installed upside down. Then I removed it. I smoked the top of the frames lengthwise.

The hive sounds changed suddenly, growing louder and more intense as the alarm sounded. I lifted the first still-bare frame and set it down against the front of the hive stand. It was not until I reached the fourth frame that I saw what I was looking for. The newly built combs were fragrant and white. There was an arc of capped honey and below it, the glistening accumulation of honey yet unsealed. The pollen I'd seen coming into the hives was neatly tucked away. Toward the center were the capped worker cells. Every time the bees climbed in my direction, I puffed more smoke in theirs and they scurried downward to fill themselves. My tolerant bees let me rip apart their home without interruption. This amazed me. Picture a giant ripping off your roof. The giant is dressed like a massive cloud millions of times your size. Maybe the giant is a god sending through a tornado and a forest fire all at once. You fear for your life, your home, your carefully accumulated savings. Your unhatched young may have to be abandoned. The giant lifts up your house clumsily, turning it over and searing it with sunlight when you like it dark. An ogre. An invader. I spoke gently to my bees, apologizing as I clumsily disrupted their lives.

In Jafenhar, quite by accident, I glimpsed the slim, golden queen. On the day I had hurriedly removed the queen cage from Thridi, I had not quite pushed the frames together so that the bees would build as bee-keepers intend. I now saw the consequences. I had unintentionally given them room for creativity. Instead of following the inspiration of the tidy pressed wax foundation, they had built up some of the frames in a fashion that suited them more, with ledges, cliffs, tunnels. All the cells were perfect hexagons, but with hills and valleys, a freer topography than the modern beehive allows. I supposed that what I ought to do was take the hive tool and "even it out." I lifted up the frame, holding it by finger and thumb. It was almost too heavy, filled with brood and honey and this artful comb. I attempted to inspect the other side. Just slowly turn the page of a book, I reminded myself. This I did. And as I did it, a weighty chunk of freeform hive cells broke off and fell across the frames. "Oh my God," I whispered. It had been beautiful stuff, filled with brood cells, perfect in every way except the beekeeper's way, and now, *thunk*! I drew a deep breath and tried to calculate how I, Godzilla, could minimize the harm. The wood and wire cages that had carried the bees from Georgia were sitting nearby. Gently setting down the frame, I placed one of these cages at a right angle to the hive. I put on my yellow-handed gloves and retrieved the fallen comb. Suspecting that a proper beekeeper would probably heave the lost brood into the woods and call it a loss, I placed the severed comb carefully atop the cage and pushed my makeshift orphanage close to the hive entrance, hoping the bees would care for their own once Godzilla and her portable hell were gone.

One the way home, I tried to calmly review what I had seen during my first inspection of the new hives but my mind was too dizzy for reflection. I found myself remembering the first time I made love: Omigosh, I thought, I've actually done it! Now what did I actually do?

4

Flowering

When you are stung . . . you were best be
packing as fast as you can: for the other
Bees smelling the ranke savor of the poyson
cast out with the sting, will come about
you as thicke as haile . . . They are like unto
incorrigible shrewes: There is no dealing
with them but by patience: though when
they sting they are sure to have the worst.

—*Charles Butler, 1623*

*A*LL THE WORLD WAS FLOWERING. EVERY DAY I prepared my sugar water, wanting, in some way, to remain necessary, but the girls merely sipped where they had supped, passing over my bland mix for the rich choice of nectars nature offered. My bees had accepted my awkwardly nailed wooden boxes as their new homes, patching little gaps with their fragrant red-orange propolis, the resin gathered from what was known, in Aristotle's day, as the tears of plants and trees and then transformed by the bees for just such purposes. They had become self-sufficient.

I returned to tinker with my orphanage. Bringing along a miniature white plastic laundry basket that had, up to then, held the lingerie bags and little things that need to be near the washing machine, I could hardly wait to get going on my scheme. Using four thumbtacks, a pair of scissors, and gardening wire, I rigged up my little contraption. Basket upside-down. Wires through the holes on four corners. Tacks on the top of the wooden cage. The solid white plastic bottom was to be the roof. The open sides would ventilate. The orphan comb was mushy from the heat of the direct sun.

There was no blanket of attending sisters. Two bees stood atop the brood and it was impossible to know why. Though their apparent indifference should have told me that the bees had already made a decision, I didn't take the hint. I wanted to save those unborn bees. They were unborn because I had been clumsy: Sentiment, tenderness tinged with dura-guilt, guided my actions. The bees ignored me. I wasn't in their way. They came and went, those self-sufficient bees, carrying this, carrying that, as I struggled in my human fashion to right my oafish influence upon the past. I should have seen that I was grasping, trying to hold on to ephemeral things. Bees know better than that.

What is the past to a bee? Where the nectar was last time? Where the sun was when the flower was found? Which eggs were laid yesterday? Which cells have already been cleaned out and made ready for honey, pollen, or an egg? The Bee Master told me that the bees I saw now, the bees working and cleaning, feeding and gathering, the workers working with all that they are, would not be the ones to eat the honey they made. They would be dead by the time it was needed. Honey belonged to the future and it is not an accident, but an effort, to make.

What do bees bring into their hives? The tears of trees. Water. Pollen. Nectar. Honeydew. In the very old days, people supposed that the pollen balls tucked so neatly into two back legs were little stones, meant as ballast against the wind. We know what it is like to walk into high wind when every step is an argument with the air and yet tiny bees most often stay their course. A marvel.

Crouching beside a hive, you do not wonder why, until very recently, mankind has allied bees with the divine: with holiness, not stinging, momentary pain. In the ancient Egyptian writings found in the Salt Magical Papyrus, we are told that "When Ra [the sun god] weeps again the water which flows from his eyes upon the ground turns into working bees." When Zeus was born, he was doomed to be devoured by his father, Kronos. His mother, Rhea, gave her child-devouring brother-husband a swaddled stone to eat instead and hid her baby away, some say, in a cave on Crete's Mount Dicte, where he was nourished by honey from sacred bees. Virgil understood honey to be a "heavenly gift" from the air, and the ancient Indian, Persian, Germanic, and Finnish people all seemed to agree that either bees or birds carried honey to the flowers from the sky, or sometimes, Paradise itself. The same bees that one cannot raise in New York City without running afoul of the law once produced what Porphyry called "the food of the Gods," the world's only natural sweetness. They were known as the Muse's birds, the link between humanity and heaven. It makes my heart sad, knowing what our spirits have lost through fear, the worship of seventeenth-century rationalism and the convenience of processed white sugar.

In our time, pollen grains are understood to be the "male sex cells in plants." Honied nectar, the ambrosial drink of the gods, is "a plant secretion rich in sugar," which frequently has "a strong odor." And the magical ferment which, Seneca speculated, bees used to convert a flower's nectar to honey, turns out to be "two chemical changes . . . brought about by the addition of enzymes from the bee's glands." Honeydew, that beautiful, evocative word, has been thoroughly analyzed. It turns out to be sugary bug poop. Sigh.

Poetry has been banished from too much of our lives. When it is discarded, the great is made small and beauty, mundane. And nothing is the better for it. Any child can tell you that the chemical composition of flour has nothing to do with the meaning of a cake.

The utter strangeness of the hive changed my eyes. It changed my seeing and made the whole farm into more. The farm was not just "a farm." It was the rainbows my daughter saw captured in the dew. It was a huge bird, the color of buffed steel, stalking through low water in the inlet. It was a great blue heron, but the name of the bird has nothing to do with him, with his life. His mind was concentrated below the surface of the water. He cared nothing for us. He lifted up his knees, walking carefully, each step considered. He hunched. He straightened up. He leaned forward, then twisted his head, his yellow beak ready. He stalked and stood and walked and stalked and stood and then—suddenly!—his head was in the water and in the very same instant out, and a wiggle at the tip of his beak said "he got his fish," and as the human mind registered "fish" he swallowed it. My daughter composed a song about this to sing to her Barbies and Cabbage Patch dolls, and this was the farm, too. The spiderwebs that bound the fast-growing leaves of one plant to another, white and blown out a bit like sheets hanging in a breeze, the apricot-colored bird running and shrieking and flying beside us: This was what my daughter called the farm's ness. It had become a part of us, and we, who had once been mere passersby, it. It changed every day and every day was changed. Certainly, we were not the same.

I could not understand why I did not see Farmer Tom at the farm as the spring advanced. I supposed that he must be coming when I was gone. The lawn sprinkler inscribed a watery arc over the sawhorse table. My daughter frolicked in the spray. The plants on his table farm grew larger every day. I wondered when these plants would be in the ground, but what did I know of that?

On Memorial Day, the Brownies marched up Main Street behind the Boy Scouts and the Boy Scouts marched up Main Street behind the Fire

Department and the Fire Department marched behind the men who had served in the wars. Some specially chosen kid read the Gettysburg Address, or at least we all assumed the kid read the Gettysburg Address, which is what we always assume although we can never hear it. Outside each of the three local churches a prayer was said and guns were fired. We drank soft drinks and waved to our kids and our neighbors' kids. Everyone applauded and May was over.

Now that it was June, my daughter had become a planet. Neptune, to be exact. In this celestial form, she was making her stage debut at the local school. I spent my spare moments cutting out cardboard stars to hang from a music room sky. My daughter rehearsed and re-rehearsed her six lines. In the aisles of the IGA supermarket the sunny voices of her classmates and their brothers and sisters and their classmates could be heard, alone, in twos, in randomly assembled quartets singing: "My Very Educated Mother Just Served Us Nine Piiiiiizzas!" until a good portion of the grown-up townspeople found that, with no conscious effort on their parts, they, too, had memorized the planet order in our solar system and could, in a pinch, have stepped into the chorus line and demonstrated their astronomical knowledge, complete with crossover kicks and Motown hand language.

On the day of my seven-year-old's splendid theatrical triumph, I realized Ruffy, our old man cat, was suffering from something other than humidity and heat. I'd blamed his mopiness on too many days of stifling weather, but something else was wrong. He wasn't joining in our celebration. That was unlike him. He wasn't nuzzling with his tattered soulmate, a once-pink stuffed bunny who had been loved and licked furless and was as real to him as we were. He wasn't curled gracefully on the bed. He was sitting with his back legs splayed awkwardly under an antique filing cabinet in the corner of my office. His eyes were sad. His blue-gray fur was dull. In the twelve years since I'd brought him home and nursed him on lamb baby food and rice, I had never seen him look as desolate as he did now. I called the vet. She suspected a bladder infection. So did I. But poor Ruffy didn't have an infection. He had an inoperable tumor. And less than a week to live.

The vet wanted him to stay the night so she could watch him. She promised to give him something for his awful pain. I returned home with empty arms and a broken heart.

My daughter and I share a distrust of minced words and euphemism. Old Ruffy was one third of our little household. I did not hesitate to tell her all I knew. My daughter considered this gleaming *eminence gris* her big brother. We held each other and grieved, stopping now and then to wish aloud that the truth would somehow go away. Though my seven-year-old daughter had never had any contact with death, she knew what to do. She dug out her favorite photographs of our happy life with Ruffy and taped them to our refrigerator door, where all important things go. Beside these she taped a message written in her best handwriting: "Ruffy We Love You."

At 5:30 next morning, before my daughter was awake, I dressed and went to the bees seeking the comfort of their vitality. I expected we'd be alone, my thousands and me. I did not expect Farmer Tom. But there he was, his face covered with stubble, his eyes bleary as if he'd just opened them. He had. He'd slept in the barn on a cot. He called to me excitedly. I waved back and continued toward my hives. The orphan comb had been entirely abandoned. The comb looked soggy and bereft. The bees had seen it for the lost cause it was and moved on, investing their energies in life, in the warmth of the hive. I dismantled my laundry basket and wire contraption and set the comb on a small clearing in the grass. But for one or two early wanderers, the hives were quiet, still at rest. The sun had not yet shown itself through the mole-gray mist.

I joined Farmer Tom in the field. "Footprints!" he declared. "I suspect bear." I stared at the vague scuff marks and a few half-dug holes. Reading *Winnie the Pooh* to my daughter had not prepared me to be much help with paw print identification. I did know, however, that bears like honey and can happily destroy a hive to get at it. Though the topic deserved more, "Mmmm" was all the comment I could muster. As full of sadness as it was, my brain had no room for bears and precautionary things that needed doing about them.

My daughter did not want to go to school and I saw no reason why she should. Instead, we would go to the vet and visit Ruffy. She sat herself at her little red-topped table and wrote a letter that was to go in Ruffy's grave. "To the people who run heaven," it read, "Ruffy goes there. Thanks." She carefully folded and taped white paper around her letter. "All important letters go in envelopes," she explained. She printed "To: Heaven" and made sure to write a return address. Then she wrote Ruffy a letter telling him how much we loved him. I signed too. She went into the yard and found a big stone to mark his grave. On this, she painted Ruffy's name. For ten of Ruffy's twelve years he had been on a dull diet. By vet's orders, he'd endured the same grim brown pellets and water every single day and night, except when he could steal a piece of pastry. He favored scones. He'd once loved tuna, so we went to the store and bought him a can of the best and some sliced ham. At the bakery, we purchased a raisin scone. This was no time for anything less than a feast.

Our kind vet had given up her sleep to tend Ruffy through the night. He looked deceptively well, well enough to hope for hope when there was none. I could see that though he was suffering, he could feel more than his pain. My daughter presented him with "Bun" the rabbit and read him the letter from us both. The vet produced some tape and we taped the letter to his wire cage. We scratched his favorite places. He managed to purr. My daughter asked exactly what was wrong. The vet explained in deep and respectful detail. My daughter asked why it could not be fixed. The vet patiently told her why not every illness has a cure, how death had a time. She took her into a pretty patch of woods behind her home and let her pick the place where she wanted Ruffy to be. My daughter gave her the letter to heaven and the stone. She described how Ruffy liked to cuddle with Bun and asked that he be buried just that way. The vet promised to do all my daughter asked, never looking at her watch, never seeming to have anything else to do but help my small child understand the incomprehensible and say good-bye. We had set up a dish next to his cage, and gently helped Ruffy to the floor. The moment I popped the lid of the canned tuna, he gathered all the strength he could find and eagerly wobbled to where I stood. He nudged my leg with his head and I set the can down beside him. With unbridled delight, he launched into his fish, lustily savoring this long-forbidden delicacy and we shared his ecstasy through our tears.

The next morning, I got a phone call and hurried north. He was leaving us. I held him. He pushed his head under my arm, purring as I stroked him and spoke softly, wanting to fill him with love. And then suddenly his face looked sharper and thinner. His life had gone. I faced his death with grief and unexpected wonder. What I saw was this: that the life force, the spirit, is a thing unto itself, and when it leaves, when it left old Ruffy, his form lost its fullness. Something had clearly gone. What was left once held his beloved self, but the self which had been present one moment had departed as if on a breeze. I understood what a soul was. It was this self. And I understood that, if energy never dies, that bit of self that was not in him was now elsewhere; where, I could not say, but I had seen it, and seen it go.

The balm for death is life. My daughter and I lit yellow candles and let them burn until they extinguished themselves. My daughter, realizing at the age of seven that death takes us all, moved, with her blankey, her pillow, her special doll, from her bed into mine. She could not be alone with mortality. She barely understood time. I understand it better, how it expands and contracts and vanishes. As soon as my little one was off to school the next morning, I was off to be with the bees.

I saw that Farmer Tom and his long-haired lady had been busily planting strawberries and the field was beginning to resemble, if not a farm, at least an ambitious garden. He had marked out a path where I was to walk in order not to stomp the seedlings. The apricot-tinted bird, which my daughter and I had looked up and decided was a killdeer, flopped and wriggled in the dirt. She flew up in the air and down again, racing along beside me, shreeing and leaping along the newly planted rows so that I had two escorts, her and the usual bee. I saw bees working inside Farmer Tom's broccoli rabe and admired their broadmindedness.

The best time to open a beehive is midday, when the sun is high and the foraging bees are afield. I was three hours early and the sky was overcast. I performed the sugar-water ritual though it was plainly unnecessary. So many rituals are. But for me, on this lonely catless day, sugar water was an offering of friendship, a greeting, my gift and apology for intruding.

Though the brevity of bees' lives make humankind look like a race of Methuselahs, for thousands of years man has watched the humming transit of these sweet manufacturers from the depths of trees and rocky caves to opened petals and sunlight, earth to heaven, heaven to earth, and seen the eternal. The belief that our human soul was embodied in a bee once covered the world like a net: in India, in ancient Greece and the Roman world, in central Europe, as far as Timor, as near as Scotland. So many bees. So many departed souls. Into the caves flew the bees. Into the underworld went those whose time on earth was over. In the old days, if you failed to tell your honeybees of a death in the family, the bees might die or desert their skeps at the loss of their master. They might fly up to heaven in search of him. Almost everywhere the honeybee was or is known, humans shared with the bees our mortal goodbyes. And here was I, compelled to do the same.

I did not suppose that I would find Ruffy's soul in their company, but rather that I would soothe my own. As I began to load up the smoker, I realized that, as it often does, distress had led to distraction. I'd forgotten my hive tool. The Bee master had been unequivocal. Rule number one: Always have your hive tool. The fire had not been kindled. My car was waiting at the far side of the field. I could have gone home to get it. I should have gone home to get it. But I didn't go home to get it. I wanted what I wanted when I wanted it, an attitude that may propel you to the highest heights of the business world, but will get you nowhere amongst bees. No sooner had I casually lifted the cover of Thridi, admiring my own daring as much as the way twelve thousand bees and one fertile queen had doubled, if not tripled the population in a month, then this young nation sacrificed one of their own to get rid of the hubristic giant. She backed her stinger right into my left pinky. I tried to flick it off. Didn't work. It would have worked if I'd had the hive tool. It would have been easy. I grabbed a stick off the ground and tried again. Too flimsy. The bee who stung me would certainly die, but her poison sac sat atop my skin and throbbed, pumping venom into my finger. As this *memento mori* delivered its heated message, I struggled to remove it. After nine tries, I succeeded, but the stinger remained, sticking out of my finger like a tiny spear. I tried to pluck it out with my thumb and forefinger, a halfway job. My pinky swelled up like a rosy sausage. It hurt in ways that I hadn't known a pinky could hurt. I might have been stung any time on any day, but on this day I had forgotten, then recalcitrantly refused, to follow rule number one. A sting can be a badge of honor or a badge of infamy. Sometimes, to paraphrase Freud, a sting is just a sting. However, sometimes a sting is a rebuke from the gods. I had been distracted, inattentive, careless, willful, and I had been reminded that no matter what my worldly cares, when at the hives, my spirit and attention must belong wholly to the bees. Or else.

I closed the hive and wandered past the animal tracks that had Farmer Tom guessing. Feeling ungratefully humbled, I promised myself to face more grim reality and at least think about bears. Before I drove off, I stopped to collect myself at the tidal pool where the heron liked to hunt. I saw two long grayish-brown objects in the water. Could it be? Farmer Tom had mentioned beavers. I turned off the engine, quietly opened the door, and stepped out. I stared in reverent silence. They were very still. There was a splash out on the river. Returning my attention to the motionless beavers, I wondered why they did not react. Another splash. I feared the beavers might be floating, asleep, maybe dead, and then I realized. They were neither floating nor dead. They were stones uncovered by the low water. I heard the splashing yet again. Looking away from the stone beavers, I saw a small sun leap from the water. Chastened, I dared not assume that something brilliant and yellow and leaping from the still, olive pool was necessarily a fish.

I left the farm for the solitude and safety of my kitchen, where, with the help of tweezers, I removed the last of my ignominious sting from my hot, wretched little finger and cursed my recent descent into nincompoopery.

I queried anyone who looked like they might know something about the manners of the local bears. Local wisdom went every which way: Bears would stay up in the hills unless they were starving; bears loved the town dump down in the valley. They would avoid human settlement. They would go anywhere food was. There was no cause to think the bears would ferret, or more properly, bear out my hives out of all the places they might possibly go for they would have to cross a very busy road. There was every probability that my hives would be a prime target because they sat beside a splendid source of water and trout.

By happy coincidence, just as I was calculating the sum of these conflicting opinions, my daughter and I received an invitation from

Red, the bespectacled ursine stranger who'd rescued me from my ignorance on day one. His pine-paneled den was filled with beekeeping paraphernalia. The cool walls of his great stone lodge were covered with the heads of the many animals he'd personally wiped from the face of the earth. Red was not only a rescuer of damsels and a tender keeper of tiny bees—several times a year he traveled the world in order to chase, find, admire, and shoot animals where they lived. He was as amiable as a man who gets to kill what he wants when he wants can be. Unconsciously combing his fingers through his thick red hair, he talked in precise detail about spoor and tracks, how you told one creature from the next, what they ate, how they passed their days and nights. His love for animals was undoubtable, as undoubtable as his desire to hasten their demise. With a paternal arm over my shoulder, he walked me out to the hives tucked into the forested edge of his property. The hives were surrounded by an electrified fence with a solar-powered battery. If the fence was on and a bear touched it, the bear got a nasty shock. He draped the fence with strips of bacon. He felt I should do this too. He loaded my trunk with a huge spool of heavy wire and some poles with holes. He lent me a coffee can full of alien tools. The rest was easy. All I had to do was buy some nitrate-laden Oscar Mayer pig meat and a solar generator and hook the whole smelly electrical gizmo up at the far side of a soggy field by a river and I wouldn't have a thing to worry about.

Except rattlesnakes.

Where Farmer Tom had not planted, which was almost everywhere, other things felt free to grow. Each day, in my chic purple wellies, I walked through the makings of a monarch butterfly's paradise. Milkweed rocketed up from the once barren soil. It seemed to grow an inch an hour. Where they could get a root in edgewise, the spiky brambles and wild grasses stretched toward the warming sun. I could no longer see the killdeer when she ran along beside me. I could no longer see my feet as I walked. A multiple choice question slithered back and forth through my mind: If a rattlesnake should leap out and chomp my leg, would A) my chic purple boots be fang-resistant, or B) the lethal fangs pierce the rubber and go right into my juicy calf leading to C) death of fright or D) death of snakebite? I do not generally waste my time anticipating the worst life has to offer, but in the case of timber rattlers, whom even the Latin name labels *horridus,* I made an exception. I have learned to feign quasi-appreciation when visits to the Audubon Center compel me to deliberately deceive my child in the name of not transmitting my ardent phobia, but I side with Adam and Eve post-banishment when it comes to loathing serpents. It was a rude shock when I learned that, by sheer chance, I had moved to the site of the largest den of timber rattlers found in the northeastern United States. Had I this nugget of information beforehand, I would never have set so much as my baby toe over the town line in the first place and would probably, therefore, not have been keeping bees. I envisioned rattlesnakes rattling away like ladies in a beauty shop, sharpening their fangs, patiently biding their lethal time, with nothing to do but bask in the sun and wait for me to saunter by. No matter that their den was across the river. Rattlesnakes swim. I decided that I would walk the same route every day so that my boots would tamp down the green and, at least along one scraggly line, I would be able to survey before I stepped.

For days and days after that, there were biblical rainstorms. The thunder rumbled without end and the lightning was so bright it did not turn night to day as thunder sometimes seems to do. It turned every night thing white, blindingly white, so that in the pre-dawn hours, the trunk of a tree or the grass lawn became so brilliant it was painful to gaze upon. When the lightning let up, I could slosh through the gummy soil with a childlike pleasure, for my purple boots were as high as my knees and the muck was only half that. I came to the bees in the capacity of the philanthropic white walking flower. My sugar-water nectar enabled my housebound girls to dine in and dine, if not well then well enough, until they could go shopping among the fragrant blooms, and so I felt happily useful once more. While it was pouring, I thought only of feeding my poor bees. Now that the sky was blue again, I felt painfully obliged to prevent the bears from feeding upon them. It was time to do my bit for bear prevention. I opened the trunk of my blue Honda, regarding the wire and the posts and the tools therein. That would suffice for the moment.

I returned to the farm, and, discovering the elusive Farmer Tom, had a chat. He had designed a logo for the farm, had a sign painted for the future farmstand, had been hard at work drafting a fund-raising business proposal to be printed on creamy paper with green ink, and had spared nothing in his effort to persuade one wealthy woman after another to bankroll his vision of the enterprise this farm could one day be. He had also managed to get most of the strawberries and some of the exotic herbs into the ground. He shared his personal philosophy about herb planting, which went: He did not plant them in straight rows but in wavy ones, going whatever way appealed to him. He had made several piles of rocks taken from the field so that, in certain sections, it looked as though druids had stopped by for a drop of celestial reckoning. There was a charm to all of this, though most of his vision of the farm was still billeted in plastic trays on sawhorse tables waiting to become acquainted with the soil. I encouraged the conversation to drift bearward. Farmer Tom seemed to weigh his words a while as if it might cause him pain to part with them. He described how he'd followed the tracks in question out past the strawberries until he'd come to a suspicious-looking hole. Not knowing why a bear would dig such a hole, he knelt down to examine it and to his surprise, his gaze was met by a dark pair of eyes and a pointed nose. And who was I to laugh when Farmer Tom's wandering bear turned out to be a burrowing turtle?

5

High Summer

I am a little Italian Bee, I am Queen of
a colony,
And I have no parents to bother my life,
I have no husband to call me "wife."
I have no taxes and no debts to pay,
Living's high cost don't come my way,
And I am as busy as busy can be
At superintending this colony.

—*Mrs. J. M. Morgan, 1918*

I HAD REORIENTED OUR TIME, OUR DAYS, OUR minds on a whim and now, two months later, it felt exactly right. The wise virgins, their few brothers, and their queens had their own beeswax to mind. Though bees do not have the kind of minds we think of as minds, they are not without language or other ample resources, many of which we lack. When they use their heads, they really do: making and secreting the wax that makes the combs that hold the honey, pollen, and brood. The DURAGILT stapled into a frame just provides the waxen foundation upon which the geometrical masterpieces must be built. In the early days of establishing a new hive, there is much secreting to be done.

On the day I brought the bees to the farm, thirty-six thousand bees had seemed an enormous number of insects. Now, as the days stretched toward the peak of summer, this vast number had easily quadrupled with no real help from me and I had learned to appreciate the unimportance of importance. What mattered, mattered, not who noticed it. The awkward business of adapting to life in our little town was less troubling when we filled the empty spaces with a pleasure of our own making, a world outside the world where we were still pointedly outsiders. We kept company as Sherlock Holmes chose to do in his retirement. And in this distinguished company, I did not look wise or wrong or right or good or bad. If I was thoughtful, from a bee's point of view, I did no harm. Sometimes I was useful. Other times, an infuriating fool.

One morning as I was delivering my junk-food nectar, I noticed a problem at Jafenhar. There was a bee that must have been caught as I slid the jar into place. Her backside was up and her head was out, but she was struggling. I feared her front legs had been pinned. I lifted the feeder a bit to free her. She must have made her cry then. Three bees hurried to the spot to help. Then more. I gently flicked her out from under the feeder. Once free, she balled up, as if in pain. By then, the spot was covered with bees. At least twenty workers rallied to the aid of their sister. One of them lifted the dying bee up from where she lay. She carried her with her small, strong, arms to the edge of the feeder and cast her over the edge. She fell to the grass. The bees had unhesitatingly come to the aid of one of their wounded but though they attempted a rescue, once her fate was clear, they did not try to cheat death of its due. They unhesitatingly removed her, as is the custom of bees with their dead, and that was all.

I and my species lack clarity in these matters. I did not dismiss what I'd seen as mere bug behavior. The bees had acted with compassion but not sentiment. A summer worker is short-lived. If she enjoys the full span of her well-spent life, when it is time, she dies with her wings ragged. The rigor of her labors is great. She and her sisters will help those who can be helped but cannot spare the time to fight a futile battle against the inevitable. Such wisdom was "an admonishment to those who would be admonished," as Sheherezade, who saved her own life with her wits, often observed.

Summer made my little kitchen too hot to cook in. We dined on our little second-story porch, and I grilled whatever had to be hot over charcoal.

Though flowers lined the railing, I rarely saw a bee. When I did, I knew it was not one of mine because we lived too far for them to commute. What amused me was that, now that I had stepped inside the gate to this new world, I knew almost certainly whose bees they were, to the extent that one can consider bees as belonging to anyone, which really, one can't at all. These bees came from a soft-spoken, gentle man in town who was socially suspect by virtue of being an openly gay single—my daughter would more accurately say "double" for one does the work of two—parent. When this fellow's bees came to call upon my potted plants, I often thought about him with a certain protective affection. I did not know the man except in passing. I had seen his tenderness toward his child, how he stopped, bending his head to listen to a small voice. I had also heard the unkind rumors, even I, who dwelled on the outermost circles of the gossip vortex. For fear a fellow parishioner's homosexuality would rub off, one of the local church pillars forbade his son to baby-sit the child. I knew that if this man felt lonely in this town, and I had every reason to assume he did, caring for the bees and sharing them with the child he had taken into his life would dull the sting of human cruelty.

When you have several thousand someones in your care, it is natural to wonder about their habits. I consulted Virgil, the man bookish bee writers love to quote, though Virgil himself relied on Aristotle and Varro, whose works I did not have to hand. He advised the beekeeper to "let green spurge-laurel bloom with thyme that smells afar and with a wealth of pungent savory and violet-beds." He warns against a yew tree near the hive and urges the beekeeper to "avoid a bog or place of stinking slime, or where the voice rebounds in hollow echo from the rocks." Long years before Virgil and in all years after him, well before agreement was reached as to what, exactly, bees drew from flowers and why, the blossoms they favored were noted. Open a book that dwells on beekeeping and you'll find a list. I might have become quite an expert if my botanical vocabulary extended beyond pretty, nice, crocus, daffodil, rose, tomato, basil, and corn.

I know lots of flowery types who can think of no greater pleasure than hours bent over a trowel, even some who spout Latin names as if they'd been dandled on the knees of Carolus Linnaeus. But though I hold with Candide in theory—"We must cultivate our garden"—the business of recreational kneeling in the dirt amongst the glistening pink earthworms has not the charm for me that stinging insects have come to hold. I prefer to cultivate those who cultivate themselves. And truth is, there are holes in my memory at the place where plant names are stored and information hardly sloshes in before it dribbles out again, despite any good intentions I might occasionally harbor.

However, I did want to know what my darling girls favored and what they did not. I bought a *Peterson Field Guide to Wildflowers* in my region. I bought *The Tree Identification Book* which is chockablock full of close-up photos of every observable aspect of treeness, except color.

Armed with these, I drove to the farm. To my surprise, I found Farmer Tom at work alongside Romeo and Juliet, the fourteen-year-old lovers he had recently employed. They were planting squash. I asked Farmer Tom about what grew on the land but if he hadn't put it in the ground himself and it wasn't pricking his ankles or giving him hives, he saw no reason to know. Romeo knew that milkweed was milkweed and complained that my bees were after him. Juliet, whose wavy hair lay loosely braided down her back, spoke softly: "They weren't her bees. You stepped on a nest."

"Damn wasps and mosquitoes," he rejoined, determined to be disgruntled about something.

In the spirit of knowing exactly what sort of milkweed my bees adored, I consulted the book. Peterson gave me eleven choices out of a possible hundred and twenty. I narrowed it down to three-ish. In the process, though, I noted *Asclepias* was the family name. Remarkably similar to Asclepius, son of Apollo and a healer so brilliant he knew all the secrets of the earth, and some that belonged to the gods.

My bees did not dwell on *Asclepias* alone, nor were they seeking to be healed. In search of sweet nectar, they sampled lemon- and cinnamon-scented basils, and certain mints. But my bees were not like some bees, who are plunked down among apple or orange trees so that they know what's on the menu. They were free to range about three miles in any direction. Whatever grew within that radius might be a possibility. It was worthwhile to know what was in bloom if I could ever match the details of the pictures with what my eyes beheld. That would educate my guesses, but they would still be guesses, for the bees might spurn an apparently suitable plant for reasons of their own. There was an untamed privet hedge directly behind the hive. Aromatic white flowers perfumed the air, calling sweetly to me and the bumblebees who feasted there. No matter. My girls, the occupants of Har, Jafenhar, and Thridi, absolutely and unanimously decreed these flowers unworthy of their attentions.

On account of beekeeping, I had, in a matter of weeks, gone through more Clorox than I had in all the rest of my life. Starting out,

I was pleased with wearing and shrinking my chic white beesuit, but it did not occur to me to wonder, Why white? as opposed to, say, a dirt-hiding denim blue? When I got far enough into beginner-ness to start to wonder, I was told the answer was very simple. Bees prefer white. Reason? If you wear white the bees won't think you're a bear and come after you. While wearing white seemed a perfectly charming idea, that bit of received wisdom astonished me. How on earth is a bee, who lives two months, at most, in the busy summer when bears are wide awake, going to think she sees a bear if a bear has not been in the bee's experience or in her sisters' or brothers' experience, or in the experience of any member of the hive at that time or in that hive's past? I could see how wearing white might blend you into the light-filled sky and give you a generally less imposing presence, but if honey-bees in the North, South, East, and West, under all sorts of circumstances, uniformly possessed a fixed bear "memory" within their race and not their education or experience, biologist Rupert Sheldrake's controversial theory of morphic fields and morphic resonance made a heck of a lot of sense to me. He says that "morphic fields, like the known fields of physics, are non–material regions of influence extending in space and continuing in time," and that "memory is inherent in nature" so that it is possible for living things to "inherit a collective memory from all previous things of their kind, however far away they were and however long ago they existed." On the other hand, when slightly pressed on the matter, the Bee Master admitted to me that he had broken this particular dress code more than once without penalty. It might be that many bees don't give a hoot about light color, but that common experience led to assumption, assumption to belief, belief to folklore, folklore to custom, and custom to presumption of fact.

Analysis of bee temperament is pretty irresistible once you have known them as anything other than something to fear or producers of that which you buy at the market and spread on your toast. It goes well beyond flowers and fashion, and is, like almost all bee-watching, an ancient pastime. Humankind was not always so detached from our fellow creatures as we are now, nor so lonely. In the time before bees were strangers, they were our teachers, with so many qualities we admired. Bees were not only the tears of a god and the souls of the dead but a gallant fighting army led by a wise king, a monastery whose inhabitants practiced temperance and chastity, a "Theatre of Politicall Flying-Insects," a model of flawless industry, the ideal commune, and a nation of Amazons, a "Feminine Monarchie." It was this last imagining that was my favorite and my daughter's, our "monarchie" being equally feminine. It heralded the awkward recognition that the head of this winged world apart, admired by all who beheld it, was, and always had been, a she. Beekeeper Charles Butler and his 1623 treatise, *The Feminine Monarchie: or The Historie of Bees* phrased the truth quite plainly: "The males heere bear no sway at all."

Not every world is a man's world, and my daughter was pleased to know it. To the degree the queen and her workers have been deeply esteemed, the drones are scoffed at and reviled. While I adore men and their company, my present circumstances allied me with the queen. My overwhelming priority was my hive. People seem to think that a woman without a man must be searching for one, but I had so much else to care about that the thought did not often occur to me. The men I met offered little beyond their maleness. And that alone was too little for me although a queen bee asks nothing else.

With seventeenth-century candor, Charles Butler frankly observed that in the realm of Her Majesty the Queen, "The Drone, which is a grosse Hive-Bee without sting, hath been always reputed a greedy lozell . . . [He is] but an idle companion, living by the sweat of other brows. For he worketh not at all, either at home or abroad, and yet spendeth as much as two laborers: you shall never find his maw without a good drop of the purest nectar. In the heat of the day, he flieth abroad, aloft, and about, and that with no small noise, as though he would doe some great act: but it is only for his pleasure, and to get him a stomach, and then returnes presently to his cheere . . . But for all this there is such necessary use of him that he may not be spared, as without whom the Bee cannot bee."

Never mind charm or lending a hand about the house, the drone exists for sex and sex alone. But at least he's clear on the point. As it happens, the pudgy drone is, in proportion to body size, very nearly the most prodigiously endowed stud of all the animal species. Only a few fleas outdo his genital splendor. He, and a few fellow "greedy lozells," couple with the virgin queen during her once-in-a-lifetime mating flight, a ritual that takes place more than twenty feet about the ground in the same exact locations year after year, a fact made more intriguing by the knowledge that no drone lives to repeat the visit that no queen makes twice. I had to wonder how anyone knew where in the air to go! Though mankind lacks no awareness of sex among salmon and snakes, rhinoceri and roses, and has long been able to describe the details, it took until 1962—two years after the introduction of the birth control pill—to solve, or at least scientifically observe, the mystery of how, exactly, bees begat. But as far as I know, how the *bees* know, we don't know.

With nothing to do in life but eat, idle, and mate, it is no wonder that the amiable Bertie Wooster's spiritual home away from home was The Drones Club. No matter that the issue was actually settled in 1623, not everyone is at home with the true nature of the male bee. A man who says he is "feeling like a drone" is not confessing to being a pampered lazybones whiling away his days as he waits for a chance to have sex with royalty. The truth about drones has been buried, the meaning has meta-morphosed into the near opposite: feeling like a drudge, laboring, as *Webster's* puts it, in a "routine, unimaginative way" at "menial, distasteful, dull, or hard work." Pity. I can easily think of more than a few people who—except for the death at ejaculation bit—would consider a real drone's life anything but distasteful.

My bees with their color vision and compound eyes saw none of this. Together, my hives made a music that sounded like a distant water-fall. I would breathe in and in and in, unable to get enough of their mixed flower perfume. Each hive smelled like the other. If there were differences, they were too subtle for me to detect. Still, it was becoming clear to me that the hives were not all the same. Each hive had its own personality. Har was rather hectic, and the bottom board was such a busy landing pad it wanted for an air traffic controller. Jafenhar cleaved soberly to the middle path in all ways; always active, but never manic. As I stood to the side and watched, it showed less pollen coming in. Thridi was rather given to understatement; a little this, a little that.

In the late afternoon on the hot, hot days, when the humidity felt crippling, those that could would gather at the portholes, carpet the

outside of the hive body at the entry ledge and under it, hanging, legs linked, in the shade, fanning so hard they looked wingless. The purple-flowered milkweed was now about waist-high, and the girls were sampling the yellow-flowered kohlrabi, which was more than my daughter was prone to do.

On the day of the summer solstice, I arrived early with my sweets. Just then, a truck pulled in behind me. Out stepped four brightly sun-burnt shorn-headed men holding strange and lethal-looking medieval contraptions. Farmer Tom was, as usual, not there. The location was too isolated to voice a cry worth anything more than an echo off the hills. So it was me and them.

They had their weapons. I had my beesuit. I pulled my ghastly plastic safari hat onto my head, signifying I meant business, and sauntered over as if confrontations with armed men were my daily bread. In what I hoped was a strong-arm-of-the-law voice I said, "May I ask what you're doing here?"

The leader, who was skinnier and sinewier and wizened-er than the rest stepped up close to me and rumbled, "Carp shootin'."

I nodded curtly, resisting the urge to ask what the hell carp shootin' was. I regarded their strange weapons. They were hybrids; crossbow, gun, and fishing reel with a nylon line attached to a spiked arrow made more lethal with the aid of two mean-looking metal barbs at the end. I attempted to smile.

The leader handed me his business card, the first line of which read "NUISANCE," with an explanatory "Wild Life Control" underneath.

"I been fishin' here for years. Caught a fifty-pounder right in this little inlet here." He told me it was illegal to carp-shoot in the river itself, because that's where the trout were, but that the inlets were OK. I wasn't going to argue.

I told him about the sun-colored splashes I'd been seeing. "Was that a carp jumping up?" I ventured.

"Not jumpin', spawnin'. You get the male, see, and he butts the female to make her eggs release out, see? She looks like she's jumpin' up, but what it is, is, she got herself a good hard whack from her mate. Then the male, he swims in a circle around her releasin' his sperm, see, fertilizin' the eggs."

I was agog.

The carp-shooter gestured toward the tabletop farm. "When the hell's he gonna put that in the ground? I heard he's twenty thousand into plants, there. And they're gonna die if he doesn't get a wiggle on."

"No doubt," I replied. And then, since it so happened I was speaking to Mr. Nuisance Wild Life Control, I raised the Bear Question.

"They got a range of five hundred miles," he told me. "If they're hungry, sure they'll find your honey. Sure."

"What about an electric fence?"

He looked distressed. "You don't wanna hurt 'em." I didn't. "Well, then, you go over t' Costco and you get one of them giant-size dog runs and put a chain link right over top and that might help."

Might. Right.

Come good or ill, I was out of the bear business. I would bow to the fates. If they chose to send me a bear, I'd take the consequences. And so would my bees. Having given up bear thwarting, I might, at any time, have driven over to Red's house and returned the wire, poles with holes, yellow plastic doohickeys, and alien tools that cluttered up the trunk of my blue Honda. But I didn't. Instead, I dragged them with me every-where, occasionally unloading them to make room for something like suitcases or groceries and then dutifully packing them back in. The reason was simple. I was ashamed to confess to Red that my bee love bore a glaring imperfection. No woman wants to hear the judgmental hiss of the branding iron as it sears her hide with the indictment of indictments: Bad Mommy. By carrying the wires and whatnot wherever I went I could avoid revealing that the only use I had for bacon and electricity was in pursuit of the perfect BLT.

For some unpatriotic reason there were no July fourth fireworks in town and those of us who cannot begin July without explosive bright lights in the night sky scattered to our favorite elsewheres to sit on blankets and bubblegum, buy glow-in-the-dark neck rings, and chant "ooooh." A week later the corn was anywhere from two to three and a half feet high, depending on the field. You could see boys on bikes pull over to the side of the road and wade into the waving stalks to pee in private and public at the same time, standing still, grinning into space, as if overcome by a grateful appreciation of nature, which, in a way, they might have been. "Yuckaputo," my daughter would exclaim in disgust. "Don't they remember they have to eat that corn?"

Hand in hand, sullen Romeo and his Juliet walked the three miles from the Capulet house to work at the farm, and at the end of the day, they walked back. They were too young to drive and there was no one to give them a ride unless I came along because the quiet Juliet had more sense than to hitch. They thought they worked hard. Farmer Tom thought they were lazy. Romeo wanted to get the languishing table farm planted. Farmer Tom didn't want the boy planting without his supervision, which was close to impossible because they were hardly ever together in the same place at the same time. Romeo thought Farmer Tom was an idiot. Farmer Tom felt the same way about him, but help, any help, good or bad, was hard to find, so he kept him on but was slow to pay. It was the perfect bad relationship.

It was four o'clock by the time I got to the field the day I saw Farmer Tom and a row of shirtless men. At work. Planting. Farmer Tom's lady with the long blond hair had turned a hose on a patch of leggy greens in black plastic trays. These chosen few had been removed from the sawhorse table and put on the ground. Now, they might actually be put *in* the ground. "It's a miracle," she whispered reverently, and I was inclined to agree. Farmer Tom had spotted some hikers coming up the Appalachian Trail and asked them if they wanted to do some planting in exchange for a place to stay, dinner, and . . . One of the shirtless men joined us and the sentence remained unfinished. "This is the bee lady," Farmer Tom's lady said.

A young man with tight golden curls and a shining bronze chest said, "Awesome."

He was perhaps the most beautiful man I had ever seen, as perfect in form as a sculpted Greek warrior in the Metropolitan Museum of Art. Only glistening, alive. And he stunk like he'd been sleeping with wet dogs in a dank cave for five weeks.

Admiring from afar seemed the best solution. I walked across the field to give the girls their teatime sugar water. I saw Farmer Tom regarding me, hand on brow, like a sea captain assessing the perilous shore, shielding his eyes for a better view. After considerable hesitation, he shuffled toward me. He weaved back and forth around his point until I learned how the unfinished sentence came to its conclusion. Farmer Tom and his lady had offered the hikers a place to stay, which was the field; dinner, which would be had at the local diner; and a shower. However, they were in no position to produce this promised, and desperately needed, shower. Because they were bivouacking at his very aged mother's house a half-hour's drive away and she was well past thinking her fifty-ish son should find a place of his own, her shower was out of bounds. It was a shower in *my* bathroom that had been tendered as one-third payment for planting.

Single mothers who live in the centers of small towns with avidly inquisitive neighbors do not accept offers to entertain six half-naked twenty-year-old boys they have never seen before without a moment's pause. So I paused. Then said "Yes." The boys would present themselves at seven, chaperoned. And I had no doubt my neighbor would part her curtains, peer across the street, and pick up her telephone at 7:01.

My daughter and I rushed through her bath and dinner. We set out towels and hotel soaps, tiny hotel shampoos, conditioners, and lotions. At last I knew why we had looted Disney World.

At the appointed hour, the curtain parted on schedule and the farmer's lady arrived minus five men. She had only the stinky Greek warrior in tow. Understanding the urgency, the others had not wanted to stop planting long enough to bathe. The warrior, too, had been reluctant to leave his work, but the farmer's lady, arguing, I suppose, for the good of mankind, had persuaded him. He showered for a very long time, and emerged smelling almost as foul as before. Only the wetness of his curls betrayed his hidden cleanliness. I offered to wash his filthy clothes. He said there was no point and refused everything but a tall glass of ice water. He told me he'd been walking since March 9. March 9 was the day I had climbed the sculptor's mountain to see his plaster elephants and descended with a new destiny. While I had been chasing and chased by that destiny to the point where I could actually be referred to as "the bee lady," he had walked fourteen hundred miles. I remarked that he must have seen some spectacular places. He replied that actually, he and his buddies, who had all traded their real names for trail names, like Shenandoah, rarely saw anything that was more than three miles in either direction from the trail. This left a whole lot out. Their goal was to walk all 2,050 miles through fourteen states, to reach the end, not to really see where they had been.

That sounded all wrong to me.

But, as Farmer Tom's lady said, in their brief detour from myopia, those boys had performed a miracle, decreasing the table farm by about half, thereby doubling the size of Farmer Tom's planted farm in a day.

The saintly generosity of the nameless hikers, and the fact that their labors had only cost Farmer Tom a half-dozen burgers, made life harder for Romeo and Juliet. They did no one-day miracles. They trimmed branches and pulled weeds and carried rocks during the hottest part of the stickiest days, Monday through Friday for an hourly wage. Farmer Tom, having entertained angels and been well aware of that fact, had nothing kind to say to these two mortals. No matter what they did, how hard they tried, the man would not be satisfied.

As July moved along, the mosquitoes staked a serious claim to the property. For their own teenage reasons, Romeo and Juliet wore cut-offs and T-shirts. They slathered their bodies with double doses of Skin-So-Soft and bug repellent, slapped at themselves and cursed a lot. There is no point visiting bees wearing bug repellent, so I covered every coverable part of myself every time I traversed the field. I gave up on beekeeper machismo and buried my hands in my canvas and yellow rubber gloves not for fear of the bees but to defend against legions of aggressive mosquitoes and vile-tempered wasps, who, unlike their cousins the honeybees, sting simply because they can.

It was hot and the bees hung like clusters of grapes off the bottom boards, which give the beehive its base. The field bees were bringing in a cargo that looked like amber-rose beads, gleaming jewels tucked in their pollen baskets. Sometimes more to the red, sometimes more to the gold, sometimes a deep golden pink, this new substance had a sparkle that made it quite different from the multicolored matte balls I had seen before. Holding a frame from Thridi, I watched a bee laden with her mysterious jewel wander a short while to an empty cell. She turned to look at it, and finding it suitable, backed herself in. She used her middle legs, or so it seemed, for they were just below my view, to meticulously bury her treasure. When she was done, she crawled out again. I was watching to see what she would do next when I felt a sudden violent jab through the leg of my beesuit. I saw no sac on the outside, so I couldn't use my hive tool to scrape it off. I was in the wrong place for tearing my clothes off, so I tucked the frame back where it belonged, gathered my things, and hobbled miserably toward the barns. As soon as I found a corner that was sheltered from the road, I stripped off my pants. There it was. Swollen big as a pomegranate, hot. No sac. No stinger. Just a nasty white blotch on my thigh that told me I'd been nailed by a wasp.

The mosquitoes wasted no time feasting on my bared flesh. Reason arrived late to the party and I scampered to my car shaking my pants and jumped in, locking the doors as if that feeble precaution would keep the nibbling bastards out.

It was then that I noticed the men working on the roof of the barn closest to the inlet. If they noticed my daring dash, they were kind enough to appear to be entirely absorbed by their hammering. I snuck off in a funk and wasn't too sad when the pearly mist sliding out of the sky and down the hills into the valley became a tickling drizzle then got heavy until the pocked olive river strained its banks for a couple of unreliable days and I could stay away.

When I returned to the field, it was with a well-packed smoker and the intention to immerse myself in a threatening haze if I heard any hum that did not belong to a bee. My ears, I realized, could now tell the difference between one insect tone and another.

The gentle bees of Har were sorely aggravated when I smoked and opened their hive. They buzzed me as I worked and it made no difference that I was the nectar carrier. They did not wish to be disturbed. They had sealed the inside cover, laying on the gummy orange propolis so thoroughly that it took several minutes of prying to get it off. I expected to find things progressing nicely, but the top hive body suddenly seemed alarmingly full. There were gorgeous frames of white-capped honey, but it was hard to get a good look. They were glued in place with propolis, and everywhere, in spite of the smoke, the bees were in high dudgeon.

I struggled, freeing the combs ungracefully. I was unready for the weight of these frames. The first slid heavily and swung through the air like a plane blown off course by a hurricane, terrifying the bees. They were angry at me and I was angry at myself for being such a clumsy piker. It is not easy being Godzilla when you aim to be a mild forest sprite. Was I surprised when I was stung on my left thumb? No. I deserved it. I tidily removed the stinger and sac with my hive tool and went on, contritely accepting the pain. But I am a lefty and contrition was not enough to keep my hands from becoming increasingly awkward. I got worse. They had had enough of my sort and let me know it. This time it was the right knuckle of my ring finger at the lower joint. I wrestled with my irritation as my penitence was tested. Not willing to pay more for my oafishness than I already had, I took refuge in the gloves. As I continued, I understood with a thrill why the girls were especially cantankerous. They'd filled up all the space they had.

My bees needed a place to make honey. It was time for the honey supers! The honey supers I didn't have, filled with frames I also didn't have. There was urgent construction to be done.

I found the two young lovers sitting side by side sulking under a tree. Romeo held a hose and was watering the yet-unplanted plants. I tried to cheer them by reporting on the wizardry of my brilliant bees but they were too blue to muster more than a twitch at the side of the mouth and a desultory mumble. They were waiting, for the second day in a row, to collect their wages from Farmer Tom. He had promised to be there at seven the night before and hadn't shown up. He'd promised to be there at two and it was now 3:30. That weekend there were three high-summer fairs, the kind with carousels and Ferris wheels and nauseating rides that turned you upside down, tractor pulls and ax-throwing, cotton candy, fried dough and corn dogs for eating, lambs, chicken and calves for judging, a chance at beer for canny minors, and dancing for all. They'd been intending to go to all three. Now Friday's had been missed and Saturday was almost over and they had no money. Farmer Tom hadn't paid them the week before. He owed them $170 apiece. Without their wages, they could go nowhere. They hadn't any money between them and their parents had nothing to lend. I hadn't thought to bring my purse so I was no help either.

Romeo gave a bitter sigh and gestured with the hose. "These're gonna die if they don't get in the ground. I offered a million times to put these plants in but he don't let me. Oh ho no. He's all over two damn counties yakkin' about his so-call farm and trying to bum money off these rich ladies who keep comin' here. He's waddlin' around here in a tie, too chickenshit to get his damn hands dirty in the dirt. And where the hell's all this damn money? Then he yells at us and the damn farm is goin' to hell and I keep tellin' him these damn plants are gonna die and he'll be out all his thousands of dollars he's got sittin' there wiltin' if I don't water 'em!"

"And we're missin' the fairs," mourned Juliet. She stretched her long legs and looked at a clump of black clouds. "Probably rain, too."

I promised to come back and pick them up if the skies broke. It seemed the least I could do.

Being confident in the relative bee-proofness I'd demonstrated at the Bee Master's and since, I did not know what to make of my left hand. My thumb was a rock, the back of my hand had become a boneless, veinless, baseball mitt and red streaks ran up my arm from the wrist. I dabbed some cooling aloe on my grotesque appendage. It swelled some more. I drove to the pharmacist who shook his head and muttered, "Bad reaction.

Benadryl." I asked if there was anything else. There was. If I had any trouble breathing that would mean I was going into anaphylactic shock and I should call an ambulance before I passed out.

Invincibility, it turns out, is like a beesuit. It comes with no warranties.

In the morning I was still breathing and my mitt, reduced by a third, was now merely a sinister red claw. I was in no position to assemble delicate frames and fill them with DURAGILT. The Bee Master agreed to come to my rescue—for an extra charge. I'd be able to bang together the boxes in a day or two. I just hoped the ladies of Har would be patient.

Not willing to test the outermost limits of my immune system, I put on my gloves the moment I lit the smoker, and I kept them on. With a puff of smoke, Thridi's quiet buzzing grew instantly, markedly louder. That is what always happens. The alarm can be heard. Still, they were gracious hostesses, offering me a treat. With the help of my frilly blue French reading glasses and their magnifying power of one, I was at last able to truly see the sight that had eluded me since my half-perceived, unreliable first glance at the Bee Master's hives: minuscule day-old eggs. A fraction of the size of a single grain of rice, they were pearly, slight, delicate apostrophes, each at about the same angle of repose, each alone in the center of a pristine cell, and visible only when the sun shone over my shoulder at just the right angle. Spotting these eggs was not merely pleasing. It was meaningful. Yes, I could now officially justify the exorbitant cost of my glasses in fact and not only in theory. But that was next to nothing. Egg-spotting meant I might finally, actually, plausibly hope to be a real beekeeper, not just a pretender dressing up in funny clothes.

For until you can recognize these, the newest signs of new life, you can never be up-to-date on whether and how well the queen mother is thriving. And this, more than anything else, you need to know. If Mama ain't thrivin', the hive ain't survivin'.

Sighting these crucial, elusive eggs would have given joy enough for any day, but in spite of my raw, itchy claw, the mosquitoes and wasps, in spite of the clammy heat and the fact that my smoker died out, I was rewarded with an unlikely thrill.

As I approached Jafenhar, some of the bees from Har circled round me, seeming irritated by my presence although I had no intention of reopening their hive. The bees of Jafenhar, however, received me calmly. I neither saw eggs nor larvae, but I saw the capped cells from which more workers would soon emerge, and then, on the second eastward frame in the top box, I saw a queen cell. Some people might have called this very bad news. I called it magnificent. I had never seen one before, but there was no mistaking it. It was unlike any other cell, capped or open. Near the very bottom of the frame it hung, full, bulbous, extended downward like a honeycombed teat. There were a few bees clambering over the cell, pawing it with their front legs. Then I saw a long, thin bee emerge. She was paler than the others, golden, with no black. The few bees worrying the top of the cell remained, but she walked straight off as if she knew exactly where to go next and when to be there. She was confident. I doubted myself despite what I had just seen. What were the odds? I wondered. Could it be? I hurried home to telephone the Bee Master. "If you saw what you say you saw, you saw it!" he barked. I wondered what he would have said about my daughter's unicorn.

"She was kind of skinny, though, for a queen."

"Well, if she was only hatched a second before, she's a virgin, so she's not all filled with eggs now is she?"

"I guess not," I mused.

"Of course not!" he corrected. "Now what you got is a supersedure. That means your old queen wasn't quite right. She mighta got hurt or she's weak and not producing enough pheromones. Now if you look up in your *ABC*," he said, referring to the *ABC and XYZ of Bee Culture*, an encyclopedia of all things apiarian, "you'll see it'll say the supersedure cells are usually hangin' down around the center of the frames and the swarm cells on the bottom, but don't worry about that. Things aren't always the way the books say. Theories are only theories based on what you think you know. Change what you know, and all of a sudden quarks are matter and the universe is different. Only the universe didn't change, did it? Just what we know. So the only way you're going to know is to be poking your nose back in that hive and seeing for yourself."

A supersedure would almost certainly mean that Jafenhar offered no honey for a crop. But then, a first-year beekeeper has no right to expect a crop anyway, since the bees have quite enough to do drawing out the DURAGILT to build their combs and preparing enough honey to survive the winter. A crop is a bonus. So, in my view, was a supersedure. It was a chance to learn.

I was so ecstatic at having witnessed the birth of a queen I forgot to ask after the urgently needed frames for Har, which was just as well because when I did, I learned that instead of making them, he'd decided to patch his roof.

I forced my claw around a hammer and whacked together two honey supers. This fifteen-minute job took about an hour and my daughter watched with rapt attention. But she wasn't interested in who was ready for what, she was interested in hammering. I embarked on a very long, authoritative lecture about how hammering was serious business and not for flibbertigibbets. She listened politely, and asked if she might make a doll couch. I had two fifteen-inch two-by-fours in my possession. And so, with no ado beyond a maternal warning to tuck her thumbs in, my daughter breezily drove her first nail. When that nail was nearly through the first piece of wood, we set it at a right angle to the second and she triumphantly joined the two. Soon she had just the couch she'd pictured in her mind. She summoned her four- and six-year-old pals from across the street to admire her craftsmanship and they hurried inside to find dolls the right size with the right kind of bendiness to sit on this fine piece of furniture. How lovely it is to be a mama and see your small daughter do fearlessly what, three months before, you hardly dared attempt. She was free of the cowardice that had shackled me. And that was no small thing.

We painted the honey supers sunflower-yellow and waited. I didn't like the way the bees dangled in five-inch-long clusters off the bottom of Har. Were they just cooling off or preparing to leave? I did a little checking, saw no queen cells, but didn't sleep easy.

When the frames were finally ready, I arrived at the bee supply store before the Bee Master. I was suited up and ready, the two sunflower-yellow honey supers piled into the backseat.

If the bees were as able as they seemed and made honey to spare, these honey supers would house my crop. At the least, they'd provide room to stretch. Honey supers are shallow, about five and a half inches top to bottom. They sit on top of a queen excluder, which sits on top of the "deeps" where the brood is laid and tended and the pollen and winter honey stored. Deeps, or deep supers, are another name for hive bodies. And the name comes not from their profundity but from the extra four inches of height.

Since you do want your honey super to hold only honey and you don't want to see your honey super become a new wing in the maternity ward, the queen can't be allowed upstairs to lay her eggs, hence the queen excluder, a metal grid with wires too close together to allow a queen to pass. Looking at this silvery grid, it seemed to me that it would exclude the workers along with Her Majesty. But what did I know? My betters had measured exactly.

Now that the ladies of Har had an upstairs in which to place their riches without Mother getting involved, their syrup days were at an end, at least until autumn, when it would be time to take the honey supers off. The hive was strong now. If Har were to have enough honey to share some with me, I wanted it to be cured nectar from their favorite local flowers and not cured white sugar from the IGA.

It fast became a damnable summer afternoon, too hot, too sticky. Farmer Tom was waiting for me when I reached my blue Honda. I wanted nothing more than to get out of my heavy cotton jumpsuit and down a gallon of iced tea but he had other plans. He wanted to complain.

He said his aged mother was getting snappish. Rather than suffer her moods, he had cleared away a small cement-walled room in one of the barns, fitted it with a padlock, and moved in. I wondered aloud if his lady liked that. "That woman can't seem to understand that a farmer has to roll out of bed right onto his field."

"It'd be pretty hard to move into a barn with no facilities."

"It has running water," he whined. "She's having an awful time with menopause. She's staying with friends. It's been hard. Nothing's worked out. The investors haven't come through, but I have plans. In the long term, I'm very optimistic. I'm going to approach every man and woman I can think of. I meet a lot of people with money. Serious money. All I have to do I sell 'em on the vision."

"What are you going to do with the plants?" The seven-, fifteen-, twenty-thousand dollar question.

"Plant 'em." He shrugged his shoulders. Even he could see that was hard to believe. "Or at least some of 'em." In any case, he wasn't going to let reality get in his way. "I'm spending time getting to know the field. Becoming intimate with the soil so that I sense its needs with my fingertips." He rubbed his thumb across his first and middle fingers in a gesture my mother used to indicate money. "I'll cut furrows in the fall. Get a large compost heap going, prepare for the next spring. I see myself as a manager—running the operation. You can hire hands to do the actual planting. It's a waste of my talent. You don't need my expertise for that. I need to maintain the overall vision. Keep my eye on the big picture. But she's impatient."

I imagined that if I were an almost farmer's almost wife, and I had to listen to executive management dreams while half the farm died on a sawhorse table two feet in the air, it wouldn't take menopause to make me cranky. "Well," I ventured, on behalf of fed-up women and unpaid teenage lovers, "Does she share your dream? Or maybe have one of her own?"

He had no answer for that, and drove off to the diner to get a bite to eat. Our chat had left me in a terrible mood and I knew how I could cure it. Instead of going home, I shoved my supplies in the car and walked back to the far side of the untended field to crouch beside the bees. To watch them fly in bearing colored loads of pollen. To watch them land and touch each other gently, inquisitively, with their antennae and their small legs. To watch the comical drones lumber about. Just simply to watch them.

In the inlet of the stone beavers and golden carp, there was a new splash. I wasn't the first to see it. The roofers saw it too, and when they saw me gaping at the water, the two of them climbed down from the roof to discuss it. What was it? They didn't know. At least we all acknowledged it was alive. There was a laying-over of twigs on the far side of the inlet. It might be a nest or a dam of some kind. As usual, my binoculars were on a peg at home, far from where they were needed. I promised to bring them and one of my new field guides back to the farm as soon as I completed an errand.

My hives were getting very heavy. I was not getting correspondingly stronger. The Bee Master had warned me that my supers might soon weigh up to ninety pounds apiece. I needed a table or I'd break my back. I took my design to the lumberyard and had them cut the pieces to order.

Feeling pretty handy, I nailed the leg bit together. They looked fine and sturdy. Just like legs. I had my three-by-three-foot sheet of wood that was meant to be the top. Now all I had to do was attach the top to these very fine legs and presto, a table. Simple enough. I stood the legs up and placed the big board on top. It looked like a table. But how, once I began banging, were the legs going to stay where I wanted them? It seemed perfectly obvious that the legs would fly out from under at the first blow.

I stood in the driveway for a very long time waiting for insight. Birds sang. Dogs barked. Cars passed. None came. So I called my daughter's best friend's daddy and sought advice. His advice was easy to follow. Load the stuff in the car and come over. Straightaway, I did. He studied my lumber and smirked. Ten minutes later I had a table with braced legs and a top. We spread an old blanket on the roof of my car and tied the table upside down on top. Slowly, slowly, I drove the seven miles to the farm. I gently untied the ropes and slid my table off the blue Honda without a scratch. Easy-peasy, as my daughter would say. Now all I had to do was get it across the field. I tried to lift it. Without the help of manly hands it was too heavy. I tried to calculate a way to tilt the table and walk it leg by leg. It seemed to me it would take forever, but it ought to work.

I never found out. Without a word from me, the chivalrous roofers dropped their work and scrambled to the ground. In minutes, the table was beside my hives in just the perfect place and the roofers and I were taking turns scrutinizing large wet mammals. None of the photos in the field guide matched exactly. Beavers with ratty tails? Muskrats? River otters? The creatures splashed merrily, dodging and flipping, cool when we were hot. Whatever they were, they were. Identify is a human preoccupation.

Which is not to say only humans have identities. There's the queen. Then the greedy drones. Then, the workers, busy as . . . The life of the chaste worker bee is one of progression from one identity to the next. There are bees that feed and tend the larvae. There are bees that attend the queen. There are housekeeping bees that clean out the cells and undertaker bees that take out the dead. There are guard bees that watch for incoming strangers and bees that patrol the inside of the hive for intruders who have crashed the gate. There are bees who help to keep the hive warm, generating heat by pumping their flight muscles, and air-conditioning bees, who, if the hive gets too hot, disgorge cooling moisture and fan it through the hive to make things just right. There are bees that pack the pollen into the hives. Bees that receive the nectar coming into the hive. Bees that are not quite ready for the field who are taking wing to learn the territory. There are scout bees that search for a new home for the old queen and her minions when a hive is due to swarm. And there are other bees, older bees, all of whom have served within the hive, who as their bodies—and thus their talents—change, are called to work outside the hives, under the sun, in the growing fields, in trees, on ponds, in the hills, gathering water, propolis, pollen, and sweet nectar. These are the bees who sup on milkweed and kohlrabi, the wordly bees with tales to tell their wandering sisters.

The ladies of Har seemed to have set up an outdoor summer kitchen along the riverside wall of the lower deep super and a sun porch along the bottom board. Those with a special bent toward industry worked on refining the raw corner edges of wood where the super fits together like a puzzle, improving my poor craftsmanship with a finish of red propolis. As I inspected their improvements, a bee with a reddish-orange cargo, stored not in the usual ball but extended, narrow, and flung back like the little wings on Hermes's heels, landed. She began to dance in a figure eight. A few steps around, then a vigorous wriggle of her behind, and then a few steps more and another wriggle. She did the dance again and again, the first time to an audience of one bee and one human, the second to an audience of two bees, and seconds later an audience of at least nine bees all faced her attentively. I knew what she was doing, but I'd read that the bee dance was performed upon the combs and had hardly expected to see it *en plein air*. This traveling player was willing to accommodate the weather and the whims of her public. I had my stings. And, my daughter pointed out, I had the blessing of Hermes, who, aside from being both messenger and major mischief, is the god of luck.

The bee dance is not merely a sexy cha-cha to entertain the troops. It is a story and a map. It is a language using symbols and abstraction, a language like and unlike ours. When a bee finds some splendid nectar, she does not keep the good news to herself. She tells her sisters in such a way that without her accompaniment they will know where to go and what to look for. Yes, she carries the scent of the flower she describes, but there is more. When doing or interpreting the dance, the bees share a linguistic assumption: straight "up" symbolizes the location of the sun, whether the sun is actually straight up or not.

When the bee dances, as mine did, in a figure eight, the angle of her dance inside the "eight" represents the angle of the flower relative to the angle of the sun. Say the bee dances at a thirty-degree angle. Her sisters know that that doesn't mean thirty degrees from the symbolic straight-up representation, but from the existing angle of the real sun in the sky. They adjust correspondingly. And what's more, the dancer, who may have flown through buffeting winds or stillness, describes the angle as a pure calculation. Thirty degrees is thirty degrees without correction for the wind. Wind changes. The location is what counts. Like a sailor who knows the stars, the bee who follows her sister's map will allow for the air's uncertainties and individually adapt her course to her knowledge of the destination. No matter that her goal is over the river and through the woods and up a mountain besides, even if a bee must take an elaborate route, the dancer's tale is her steady guide and she is not fooled.

How far must she fly? The dancer tells her. That vigorous wriggle symbolically describes distance. If she is doing a figure eight, and not another dance which is known as the round dance, the source is not in the immediate vicinity of the hive. Karl von Frisch, who won the Nobel Prize for his patient, painstaking work toward deciphering the basics of this astonishing nonhuman language observed that "the distance is indicated in a rather exact manner by the number of turns in the wagging dance which are made in a given time." Faster dancing, closer flowers. Slower, farther to go. What von Frisch did not discover was that as the dancers danced, they sang. The dancer's wings, beating at 250 cycles per second, create a vibrational hum in the musical key of B. She only accompanies herself in this way during the part of the dance when she wriggles her directions along the telling angle, adding further detail to the map.

How does she know how far she has gone? The most recent human exploration of the "calibration" of a bee's "odometer"—a choice of words that falsely implies that bees are not living creatures, but machines— involved an international collaboration of scientists using a six-meter-long tunnel designed to create "the illusion of a lot of territory speeding past." According to them, our best guess is that the articulate dancer measures distance by the "flow of visual images," and that "an image moving about eighteen degrees across the bee's eye triggers one millisecond of dancing."

Bees can distinguish between shades of white better than a finicky interior decorator, and though their eyes favor shades from yellow to blue, they make great use of their ability to distinguish ultraviolet, which helps the girls detect nectar worth dancing about in the first place.

Considering the difficulty we humans have understanding each other even in our most intimate associations, that amongst ourselves we have close to seven thousand living languages and have yet to decipher all the known languages that belong to our own human past, the extensive effort that has gone into understanding however little we do of what bees tell each other gives me hope I might not otherwise have. The more experience I have with humankind, the more I can't help believing that stubborn anthrocentricity is a ruinous habit.

There are more described species of bees—about twenty-five thousand—than there are amphibians, reptiles, birds, and mammals put together. And yet, I was surprised to learn that not only did my bees describe their world to one another, they did it slightly differently from race to race. Bees not only have language, they have dialects. According to von Frisch, my own good-natured Italians, were they to tell their precious story to Indian bees or dwarf bees or even other Europeans, would encounter a problem they, unlike two-legged linguists, would never find at home. Misunderstanding.

In our house the month of August began at five in the morning. My daughter's top front tooth had been dangling disgustingly for days. It hung by one pink fibrous thread. She took pleasure in seeing me wince as she twisted it from front to back with her tongue. At last, with the dawn of the last true summer month, it took leave of her mouth. Despite the early hour, there was much celebrating, and the rest of my daughter's day was spent waiting for bedtime so that she could place her lost tooth under her pillow and wait for the tooth fairy to make her visit.

By eight that evening, my daughter's exuberance was tinged with worry. Low in the sky, where the moon was expected to be, hung a huge and glowing apricot, nearly, but not completely spherical, slightly fuzzy at the edges. It was hard to look away. That was the problem. My little one feared that the moon, being so fine and entrancingly beautiful on this important night, would distract the tooth fairy from her rounds. Fortunately, this did not happen.

The next day was a perfect bee day. Near the barns, some creature had made a feast of a gray bird. The feathers were scattered in a heap and there was no meat left on the few bones remaining. I saw no blood. Suddenly thinking of sharp-toothed creatures who mangle birds led to suddenly thinking of timber rattlers, the thought of which deeply rattled me so that I was suddenly nervous about crossing the field to my hives. The lavender milkweed and pink-fingered grasses were up to my head. My path, which was only as wide as my purple wellies, remained only because I trod it purposefully, tamping down the growing and green in the same place in the same way nearly every day.

I thought of this weedy domain as an enchanted place, even with my fear of snakes. It was alive in more ways than I could ever have imagined before I cared for bees. From the road it looked like very high grass and that was all. But within, every step meant another colored blossom, a crunchy pod, a minuscule bud, an iridescent dragonfly testing a furry leaf, a moth dancing on a breeze, a monarch flexing her wings, a killdeer exclaiming, her companion's reply, catbirds, cats, a ladybug on your hand, the prick of a thistle slowing your step, the gummy white ooze of milkweed milk, the aroma of privet, the brandy smell of ripening honey, the stench of mud, the weight of a bumblebee defying gravity, brown ants marching on parade, black ants shouldering a worthy burden over ant mountains for ant miles, a rusty beetle carrying a honeybee to an unknown place, honeybees alighting, honeybees aloft, everywhere honeybees, altos in an integrated insect chorus intoning like Buddhist monks at prayer while the birds scattered melody and the spiders silently strung their filament from stalk to stalk catching light and turning it into rainbows, the changing scent of the fast slow green brown choppy mirrorslick river, contemplative herons, the commentary of crows, swallows dipping suddenly down, all in all a world, whole and vital.

The pollen coming into the hives was more varied than it had been all summer; red-amber, marigold-yellow, black, and a bright, silvery green. While the first honey super atop Har was already completely drawn out and Thridi was progressing likewise, though not as fast, Jafenhar seemed to show the slowing effects of the supersedure. The new queen, beginning late in the season, had more to do in less time. A bee drifted in, a bee drifted out. Still, I was encouraged by their zest for sugar water. They were drinking a Ball jar a day. I felt a special attachment to the bees of Jafenhar. I wanted the new regime to thrive. This time, my attentions could make the difference between weakness and strength as the winter months approached.

Farmer Tom was wretched. He telephoned me in the night to ask why I had my middle name. What he really wanted to say was that after fifteen years by his side, his lady had left him and gone far away. "I'm no *good* alone!" he wailed. I padded into the kitchen to make a cup of tea, knowing we were going to be talking a long while. The lady and the

farm were intertwined, he explained. In his mind, they were nearly one and therefore, if his love life had not been in disarray, the farm would not have gone haywire. In short, it was all her fault. His lady was bereft of faith. She should have stuck it out. "I'm not good *alone*," he cried a second time, changing his emphasis.

"Fifteen years is a long time," I began. My marriage was half that length. "She wanted children, didn't she?"

"Oh, yes. She loved them and they loved her."

"It's funny that you two never married."

"I'm very picky," he replied.

The angels weep over lost time. It was a saying I had just learned from my child who had learned it from another child. Children knew. Bees knew. Farmer Tom wouldn't understand.

While the angels were weeping, I met a man who was weak on English and strong on charm. He was a professional athlete from Eastern Europe. He'd left his wife and son behind for the summer. About once a week they spoke on the telephone at strange hours of the day or night. Someone was always half-asleep. It was hard to coordinate Czech time with ours. He seemed drawn to our tiny family and though he would not come to the hives, he joined us when we picked blueberries on a nearby hill. In his country, he told us wistfully, at his country house, blueberries were got by going into the woods. It was there he had gone when he was four, the year of the Prague Spring. Though he was a small boy, he remembered the troops that put an end to that brief flowering of what Dubcek called "socialism with a human face."

Amongst the bushes was a well-formed wasp's nest. I moved away, remarking that I would not worry about being attacked by bees but wasps

were another story. He assured me that honeybees were just as danger-
ous. They had troops that protected the hives. "The male bees do this,"
he told me. My knowledgeable daughter contradicted him. The bees that
defended the hives were female. He shook his head at such silliness. I told
him she was correct, scientifically correct. "Bosh!" he declared with obsti-
nate certainty, "Troopers are men." We weren't going to get anywhere.

That night, my daughter packed a nightie and a toothbrush and went
to visit a playmate. I made sweet and pungent barbecued ribs and but-
tered corn, a colorful salad with a pale pink vinaigrette, and of course, a
fresh blueberry pie, seasoned with brown sugar, nutmeg, and lemon.

The next morning, I savored the memory of a perfect summer meal
and the pleasure of succumbing to another of the foreigner's strongly held
opinions. I surprised myself. Maybe it was the only language in which we
could converse together fluently. Who knows? But the fact that his wife
and child happened to be an ocean or two away didn't make him any less
married. Amused, bemused, confused, I was ripe for one of those little
jokes that fall like a pebble in a pond. One of those magnificent barbs
brought to you by the Universal Wit. For what, after a summer of careful footsteps and ineluctable anxiety, should this instant cross my consci-
entiously tamped-down path? A slithering streak. A snake in the grass.

I leapt up into the air and landed laughing. Served me right!

Ruffy the cat had been dead for nearly two months and my daughter
mourned him daily. The weight of her loss added a pinch of sadness to
even her brightest days. Though she emphasized that old Ruff could
never be replaced, she began to wonder if maybe it was time to get a

kitten. As soon as she was ready, so was I. And so, the loopy, loud-mouthed Athena, misnamed after my daughter's favorite goddess, entered our lives. From the start, Athena recognized that almost every-thing you wanted to know about a person could be found in the face. She gazed intently into my daughter's eyes and my daughter gazed back with love. Athena slept draped over my neck and yelled when she was relocated, which she always was. Temporarily. When scolded, Athena scolded back, but she kept her claws out of my grandmother's newly repaired Empire couch. That, and a decent night's sleep, was all I asked. The sleep, I did without.

All my babies were happy now.

Fueled by caffeine, I dragged myself to the field. Everything was as it should be. The river was the color of rosemary. Har's hive bodies were packed with eggs and larvae. Their winter stores were ample and several frames of honey in the sunflower-yellow honey supers were capped and ready for harvest. I would have a crop! Thridi had not this excess, but it looked to have enough. Enough was all I could ask. Even Jafenhar seemed to be heartier, neglecting my heavy syrup as more and more field bees discovered the bolting herbs in the garden, the newly abundant goldenrod in the field, and whatever nectar might be found buried in the sweet petals of flowers still fresh in late summer, blossoming along the road and in places unknown to me.

Then I went away for a day and a half. When I returned the field had been shorn. As I drove toward my home, not intending to stop, I saw it. The enchanted place was stubble. My lungs tightened. It was hard to breathe. And there was Farmer Tom wielding a weed whacker, demolishing the rich unruly kingdom he had allowed to flourish in the unfarmed land. The high grasses were whacked. The milkweed was whacked. The goldenrod was whacked. The clover was whacked. His crops, such as they were, were untouched, but the world around it, whacked, the world of the ants and the carpenter bees, the wasps, the bumblebees, the monarchs and moths, the world of my honeybees and the dragonflies, the turtles, the foxes, the cats, the birds, where my cheek might be brushed by a wing, my ears startled by a buzz, my beesuit painted with pollen, all wantonly decimated. Whacked.

But why?

I drove home, unloaded my luggage, and tried to collect myself. It would not do to come at him in a fury.

When I thought I could speak without a tremor in my voice, I went looking for a reason. Farmer Tom's feeble accounting magnified the tragedy all these living things, my daily companions, had just endured.

A woman I have never met, and never hope to, belonged to the relevant Land Trust committee. This woman was very wealthy and always had been. She had never married and was very, very, very tidy. This tidy woman loved nature. In its place. The right sort of nature attracted the right sort of people and kept the wrong sort of people in check. That, apparently, was why she had joined the Land Trust. She knew how nature ought to be. Well-trimmed and ordered. Plants should be in rows and pretty to look at from afar. She was irked by the poor performance of the carefully chosen clean-handed organic farmer. His vision had enchanted her. His farm was an embarrassment. This was not the rural charm she had in mind.

If she had waited a month or two, autumn would have changed the scene. Most of the life that now depended on that messy grass for sustenance would be taking leave or taking shelter from the coming of the cold, barren time. The grass would have died, the monarch pupae would have emerged to begin their long migration. The bees would have returned to the hive or their nests in the wood or the ground. The birds knew what to do. But she was not the sort of woman who waited. She gave an order. Now. Farmer Tom did not resist. His crops were coming to their peak without the means for harvest. His tomatoes and squash were truly dying on their vines. The table farm was dying on the table. His lady had left him. He was no good alone. The nature-loving Fury had screeched in his ear. He scoured the field with a weed whacker. A habitat was destroyed.

6

A Sweet Harvest

The oldest honey I have seen is in the
Agricultural Museum at Dokki in Egypt,
where two honey pots from
New Kingdom tombs (c. 1400 B.C.)
still have their contents in them.

—*Eva Crane*

*S*EPTEMBER ARRIVED WITH A LONG LIST OF things to do. Anticipating the start of school barely registered. Summer would end not when the calendar said it did but with attendance at an annual agricultural fair that went on for acres and days. When you'd surveyed the world from the top of a Ferris wheel and walked the haunted house so many times you could anticipate the creaks, had your handbag nibbled by goats and licked fresh ice cream in the cow barn while the cows nursed and mooed and pooped around you, when you could not decide what ached more, your tummy or your feet and you struggled for sleep while your mind rehearsed a country ballad you'd heard twenty times that day, only then would summer give way to school shoes and pencils. And the fair was a week away.

Meanwhile, I had received my latest briefing from the Bee Master. It was time to harvest what honey the bees of Har had cured in the sunflower-yellow supers, and perhaps reap a tiny bit of surplus from the stores of the girls in Thridi. My bees had been generous. I had a crop! I needed a plan.

When I told the Bee Master where the hives were located, he moaned and scolded me for not putting my bees near the road. As I contemplated hoisting honey, heavy honey in wooden boxes, and hauling it all the way to the barns, I began to comprehend why the picturesque set-

ting of my hives at the far side of a wide field beneath the green hills by the peaceful riverside was nothing if not the cockamamie conceit of a romantic amateur. A full honey super would weigh half again as much as my daughter. My Honda was hardly an off-road vehicle.

"You need a truck," the Bee Master decreed. "You're in agriculture now."

I agreed. I knew just what kind of a truck it should be. One with not only the usual front seat but a roomy backseat that would be comfortable for my daughter, her cat, and their toys. One in a girly offbeat color. Maybe even purple. An unmistakably, absolutely creamy, feminine truck. But in only two hundred and thirty-one dollars I would own my blue Honda free and clear. A truck was not in the budget. I would have to buy a cart.

It was time to harvest the honey not only because autumn was nearing. It was time to harvest the honey because it needed to be off the hives in order to medicate the bees against the horrible mites and microbes that make it especially hard to be a bee these days. *Varroa* mites, tracheal mites, and Chalkbrood are just three of the nasty invaders that have diminished the wild honeybee population and threatened hive bees in recent years. Unless you are a criminal misanthrope, you positively don't want to be mixing bee drugs with "people" honey, so you time the harvest to more than just the end of flowers for the year. Like humans, bees must take their medicine at the right time in the right way. The September batch had to be administered and sit around a good long time before the cold set in. The Bee Master sold me Apistan and Terramycin and Fumidil and menthol, warned me about toxicity, and sent me home to do chemistry. I understand how to mix fruits and spices, not pharmaceuticals. By the time I got home, I'd forgotten everything he told me.

Deciding to put the cart before the drugs, I wandered over to the True Value. As usual my buddy announced my arrival to his dog, "Here she comes!" Even the dog seemed prepared to be entertained. "What're you up to this time?" I explained. He beckoned. I followed him down a row of wheelbarrows. "Here. The cart of your dreams." I beheld the fabulous AgriFab. It wasn't pretty, but it was built flat. The other carts curved at the bottom and would never do. "Let's see if we can get her into the trunk." The cart was a she. She would not fit. We tried to twist her into the backseat. Impossible. "Now don't you worry," he said. "You come back in a couple days. I'll customize her for ya." A custom AgriFab. My idea of ultimate chic, with or without monogram. He had me and he knew it. Even the dog knew it.

The night before the last big summer fair, I crawled into bed with my daughter and Camus's *The Plague*. The rats had come out and were dying on the streets with a flower of blood on their mouths. Dr. Rieux had admitted to himself that plague had taken hold in the unexceptional commercial town of Oran. And there was no anti–plague serum. Not yet. When would it arrive? I stopped reading only because my eyes would no longer stay open.

A few hours before dawn, I awoke feeling like one of Camus's dying rats. I lay there wondering what possessed me. It was warm and I was very cold. I suspected I had a fever. I walked to the bathroom with some difficulty. I found the Thermoscan. 103.7 degrees. My stomach began to turn. I knelt over the toilet and let whatever was going to happen, happen. The next thing I knew, I was having frantic dreams, dreams from which I was trying to escape, sweaty, terrifying dreams. I opened my eyes. I was lying on the tiled bathroom floor. I had no idea how long I'd been there. I had fainted.

Frightened, I made my way back to my bed. The instant the hour became decent, I telephoned my mother and asked her if she might come down. And then I fainted again.

I called my doctor. It was Labor Day. He had to be paged. My temperature rose. The doctor was astonished I was not more delirious. I told him it was sheer will.

My mother and stepfather had to make the two-and-a-half-hour drive to my house. I do not know how many times I lost consciousness as I waited for the sound of their car. I do not know how many times I lost consciousness that week. My tongue looked like an ice field, dead white. I had a rash up my arms, a rash that developed and grew until I itched inside and outside and was flushed all the time.

It was scarlet fever. How many times had I watched *Little Women*, weeping as Jo's beloved sister Beth came to the brink of death, and then after a tenuous reprieve, a brave soliloquy, and a final glance at two twittering birds, sailed over it? Somehow, as I drifted in and out of awareness, my mother and stepfather made daily life continue. And unlike Beth, I had Ceften. According to the little folded sheet enclosed in the package, a single thousand-milligram dose would clear up something oxymoronically called "uncomplicated gonorrhea." I took a thousand milligrams every day for two weeks and my fever still never dropped below 100 degrees. I was weak beyond imagining but I knew I had begun to recover when I started to feel impatient. Enough with this anachronistic literary disease! Time to pick up the pace. The angels were weeping. The bees couldn't wait.

I had a plan. As soon as I bid my parents a grateful good-bye, I vowed to begin it. At risk of ugly consequences, I had to go slow. I would take a week to do a day's work, but I would get it done.

My buddy telephoned to announce that my fabulous AgriFab cart was ready when I was, but my first trip was to visit my bees. I'd missed them. It was a short drive, but it seemed exciting and new. Beside the corn, golden grass waved in the slight breeze. The stalks made a gold ripple above the earth, as if they had absorbed the sun and become a stretch of small torches, all tiny but alight. I saw Farmer Tom and asked about the waving gold. "What do you think this is?"

With an indifferent shrug he answered, "Some kind of grass."

The bees had become enthusiasts for purple basil. The hives were wealthy now, rich with their winter stores. The ladies of Har relentlessly battled two paper wasps trying to enter the hive, they fought a black and white bald-faced hornet and nudged a bumblebee off the premises. If I was going to rob Fort Knox, I was going to need a well-stoked smoker.

On my way back across the stubble, I spotted a plant I didn't remember. It was close to the ground and had escaped the whacking. Small green-striped spheres three quarters of an inch around covered the vine like shiny fairy-sized watermelons. "What is this?" I asked Farmer Tom, assuming he had planted it.

"Some nasty weed with prickers," he said, and though he was smiling and his tone was light, I felt his heart had completely turned against this land.

My buddy at the True Value proudly displayed his ingenuity. He had taken off whatever had permanently joined the wheels to the body of the AgriFab. He had filed the axle and reattached the wheels with hardware that resembled voluptuous hairpins. The wheels were now removable. And, on account of a few unscrewed screws, the handle could now fold into the cart. He gave me a lesson on how to make the cart whole. He was so pleased with the way he'd made it all work that he gave it to me several times in sequence, to the point where, for once, I actually remembered what to do. I drove straight to the field and dropped off the body near the table farm. But only the body. That way, AgriFab would be safe from covetous custom-cart snatchers who might be lurking about. A cart without wheels is no temptation.

My daughter was eager to assist in the harvest. We went north and west over the state line to an old-style department store that sold everything anyone ever used to need. They had farm clothes on card tables in the basement. We found a heavy cotton jumpsuit with snaps and zippers and with the help of a bugproof mesh hood, child-sized gardening gloves and some duct tape, she was set. More set than I was. Any exertion promised to send me tumbling back into bed, exhausted.

The Bee Master sold me three small white plastic ovals for more than anything that size and material ought to cost. He called these "bee escapes." In theory, these little items fit into the oval-shaped hole in the top of the inner cover. You took off the honey supers, lay them to the side, and lay your oval-filled inner cover atop the hive bodies. You then put the honey supers back on top of that and all the bees thoughtfully marched down through some well-placed holes in the white plastic to their winter homes in the deeps. Given a few days so everyone has a chance to wriggle out, the honey supers would then be virtually bee-free and ready to harvest.

Sounded good to me. Jafenhar had no honey supers and thus needed no bee escapees. I decided to begin with Thridi. My daughter and I suited up and I grabbed a sheet of yellowing newspaper for the smoker, noticing that what was about to burn was George Wallace's obituary. Looking down into the deep body of the hive, I could see that the frames were jammed tight with Thridi's winter honey stores. I inserted the white plastic doohickey into the oval hole of the inner cover, placed it neatly under the queen excluder, and called it a day's work done.

My daughter fed sugar syrup to the ladies of Jafenhar and we were off across the field, filling the crown of my pith helmet with unharvested tomatoes and ripe ground cherries wrapped in tan papery lanterns.

Before I mixed up the medicine, I purchased enough latex gloves to outfit the Centers for Disease Control in Atlanta. I went over the tiny print on the directions several times. Having never got past severing frog heads in high school, I was deficient in all knowledge of chemistry. Not wanting to make a fatal debut, I attempted this new enterprise with a fastidiousness better suited to the concoction of delicate explosives. I cleaned out the sink and cleared away the dishes and separated out the measuring spoons, cups, and kitchen implements that would be forevermore set aside for the bees. Then I heated and cooled and measured and mixed and hoped for something less than the Nobel Prize but better than disaster.

Outfitted with drugs and strips and patties, three sets of condomesque gloves in addition to the usual, and pewter-gray industrial strength garbage bags, we returned to the farm.

I opened Thridi to find that a renegade clique of fifty or a hundred bees had boycotted the bee escape. Damn them, I thought irritably, feeling ruthless now that their more obliging sisters and brothers added up to around sixty thousand per hive. I had 180,000 bees to tend to and I was in no mood to humor rebel holdouts. I had the bee medicine laid out upon the table. I had but a smidgen of energy and strength. This was the *diem* and I was determined to *carpe* it. Removing the first sunflower-yellow honey super, I swept it, inhabitants and all, into the gray plastic bag and duct-taped it closed. Then I did the same with the second. Revolution quelled. Next, medicine in. Buzzing bags resting flat in my arms, I and my trooper marched to the car. By the time we got home, most of the bees who had shunned the bee escape had found their way out of a taped shut 2.5 ml plastic bag and were touring the interior of the Honda. I opened the windows, hoping the girls would find their way to the local flora. They were too far from their own hives to get home. Perhaps they would find their way to the hives of the beekeeping single dad. Perhaps they would be able to sneak past the sentries and into the hive. I regretted my churlishness. Perhaps they would drift around and die. I promised the bees that remained in the car that if they were there in the morning, I would take them back to the field.

When I emerged from the house, morning tea still in hand, there were plenty waiting to take me up on my offer. It was time to visit Har. In my weak and feverish state, simply putting the bee escape where it needed to be would be enough challenge for one day. I consulted with the Bee Master who tactfully guided me toward the enlightening realization that by placing the tiny escape hole in the inner cover and the inner cover itself under the narrowly spaced metal wires that made up the queen excluder, I had virtually rendered the bee escape useless. The honey supers at Har were thick with bees. This time, I'd best get it right.

I was resolute. I was ready. The weather was not. It rained until the first day of fall. All of a sudden there were leaves on the ground. Small comfort that they had barely begun to lose their green. The farmers had cut their cow corn nearly two weeks before. I was running out of time. The day was bracing and bright. I waited until the sun was high and set out for Har. With everything except the bee escape. So I began again.

I had already moved my table to the right spot. I lay my equipment out with Girl Scout precision. Everything was in order and in its place. I stuffed the smoker: the paper, the twigs, the twine, the fresh green grass. A lovely orange flame danced upward and I applauded my accomplishment. At the end of the season, I could finally build a perfect fire. I closed the top. All the ingredients fit as they should, except a few inconsequential wisps of grass. The smoke came out in a cool and steady stream. I sealed myself into my beesuit and taped the gap where the zipper joined at my neck. I was as stingproof as I was going to get.

Whammo!

It wasn't a bee. It wasn't a wasp. The tip of my middle finger touched the hot metal smoker that encased my beautiful fire. The finger lost all color in an instant. It looked dead. Desperate with pain, I shoved it into the spout of a quart jar of water I kept at the hive. The finger went in. It wouldn't come out. It was stuck at the knuckle. I tilted the jar so I could soothe the burn while I went through the excruciating task of wriggling my braised appendage out of the stiff plastic spout. Gazing at this raw, throbbing wand of yellowed ugliness that was now a part of my hand, I thought, So much for being prepared. It's the thing you don't expect that nails you. I had no intention of leaving. I covered the worst part of the burn with one of my daughter's neon Band-Aids as a cushion against the pain and slid on the gloves.

I puffed smoke toward the entrance to let the girls know there was a crisis. The humming grew louder. I removed the outer cover and puffed toward the oval at the center of the inner cover, the one that would hold the bee escape. With hive tool in hand I began scraping away the red propolis which had glued the inner cover to the top honey super. I lodged the hive tool under the cover to pry it off. And the cover broke. One long strip of wood came right off. I wanted to cry. Instead, I dug in my bag and found my upholstery hammer and three ¾-inch nails. I stuck the slat back on the cover, which sat right on the hive and, I, who hardly knew how to find a repairman much less be one, simply put the damn thing back together.

I had known that lifting the honey supers would be too difficult for me in my feeble state, so I'd brought along an extra deep super. I inserted the bee escape in the inner cover and set it to the side. One by one, I unloaded the heavy frames of wax-capped honey into the extra deep. It was a preposterous and time-consuming procedure. But it was my best, my only, scheme. It worked. In two and a half hours, I'd accomplished a ten-minute job and I was proud.

I drifted across the field, gathering gooseberries and flowers Farmer Tom had said he hadn't time to harvest. With these, my daughter and I would celebrate.

It was best to give the bee escape a couple days to work, so I figured the harvest for a Friday. A few days earlier Red had called and offered to help. Not wanting to become the Blanche DuBois of the beekeeping world, I'd demurred. I'd planned to haul the honey across the field in my custom AgriFab, all by myself. I now understood that, with all the loading and unloading the harvest would be required both at hive and Honda, not to mention the pushing of a heavy cart back and forth over the sharp stubs of shorn plants across the rutted earth, I might be wise to give way to my inner Southern belle. I called Red back and accepted the gallant offer I'd so lately refused.

That night, there was a frost. Red said we couldn't wait the extra day, the harvest had to be now whether or not the bees had finished escaping. I felt a twinge. Summer had truly ended. Too soon. I met Red at his office. He was dapper in a tweed jacket, dark wool trousers, and spit-polished shoes. He suggested I follow him to his house so he could get his bee duds. I dashed into his bathroom and got into my suit. I threw my jeans, sweater, and purse into the Honda. Red removed his jacket and tie, slid an abbreviated beekeeping pullover over his head and considered his transformation complete. The polished oxfords would do. He wanted to take his truck, so I threw my smoker, garbage bags, and hive tool into the back.

Red piloted the four-wheel-drive truck straight over the field, stopping ten feet away from the hives. He unlatched the tailgate. Red wanted to do the job swiftly and powerfully. I wanted to linger in this bittersweet moment saying good-bye to the summer with my bees and their fellow creatures in the wayward field, and then good-bye again, to my routine, the familiar trek that would soon become unnecessary, to warmth and time and a season passed. I wanted to savor and cherish what was for some the pinnacle of the beekeeper's year, even if it wasn't my personal pinnacle, even if my heart preferred the everyday tending of 180,000 lives to any grand hurrah. In short, I wanted to delay the inevitable. But then, without so much as a match to the smoker, my bearlike friend hoisted first one super, then the other, up and off the hive and into the garbage bags I held open. He took a moment to admire the shining dung of a gray fox, then we rode off leaving my dear AgriFab sitting unassembled and unused under its blue tarp by the barn. Harvest over.

Red carried the honey supers, a hundred-some pounds of dead weight and buzzing bees, up a flight of stairs, through my hallway, and out onto the second-floor balcony, which constituted my only outdoor space. He neatly stacked boxes one and two. I tucked a long sheet of clear yellow plastic around my honey crop.

It seemed like a good idea at the time.

It is a truth universally acknowledged that a beekeeper in possession of several frames loaded with capped fragrant comb must be in want of the honey. The question is how to get at it. Since about the time of the Civil War, unless a beekeeper cultivates honey to be sold in the comb, which requires different frames from the outset, beekeepers who work with moveable frames in hives like mine have favored separating honey from its perfectly constructed container with the aid of a hot knife and centrifugal force. The hot knife cuts away the delicate wax seal the bees have carefully put in place. The thing that then extracts the honey by spinning the frames around like astronauts in training is, to no one's surprise, called the extractor.

This much I knew: I could never be satisfied with letting someone else extract our honey. After a summer of dedication, it would pain me to be handed a bucket containing the end result. Nor did I want to stand to the side and watch. My daughter and I agreed. Nothing less than doing the job ourselves would make us happy.

But just because I drained our bank account and resorted to mealtime economies to acquire one rotund motorized stainless-steel extractor and a very large hollow knife with a cord did not mean I was ready to shove two plugs in the wall and get honey. There were preparations to make. The extractor had to be bolted down somewhere. Since I had nowhere permanent, a temporary base had to be devised. I turned to the Bee Master for advice. A board slightly larger than the base of the extractor. Four 3-inch clamps from True Value. Screws, washers, bolts. The Bee Master and his screw gun. Presto!

For want of a honey house, we used the kitchen. My daughter and I covered the chairs, the toaster, the table, the floor with translucent white plastic sheeting. Our cheery kitchen soon looked like the first day of clean-up after an environmental disaster. Screwed to its base, clamped to the wooden kitchen table with its whimsically carved legs, the extractor stood five feet in the air.

As I washed my daughter's hair that evening, we made our plans. I would wait for dark, when the bees were logey, and then brush them off the top super. Then, at five the next morning, well before we'd need to think about dressing for school, we'd spin out our first batch of honey.

My daughter went to bed early, skipping her bedtime story in favor of extra sleep. I suited up as if I were going to the field, and stepped out onto the balcony with my bee brush in hand. The extractor would hold six frames at a time. I set my spare super in the house, near the balcony door. I lifted the yellow plastic blanket and opened the garbage bag that enclosed the top honey super. Vigorously applying the nylon bristles to one frame at a time, I brushed off the bees, dropped the frames in the waiting super, shutting the door sharply each time to thwart any attempt by the bees to return to their honey. I tucked the remaining four frames into their super, inside the garbage bag and under the yellow plastic blanket, and then tucked myself into bed under pale damask sheets and down.

I was up before the alarm, anticipating the delight of spinning our crop into gold. My daughter joined me in the kitchen. It was too dark. The light over the kitchen table hadn't worked for ages owing to some fire hazard wiring problem my landlady hoped would go away on its own. Since we were used to eating by candlelight, the problem of seeing what we were doing had not occurred to me. At the first light of day, I called for silence. The silence was not reverential. It was a safety matter. The uncapping knife was nothing if not a murder weapon. You plugged it in, waited for it to get hot enough to make wax sizzle, and then pressed it against the bottom of a frame, and cut upward toward the heart, aiming for the face, keeping fingers and thumbs tucked so they would neither be burned nor severed. The knife was heavy, the slicing tricky, and an aura of heat surrounded the double-edged blades that came to a point just perfect for poking out an eye. I aimed to cut the wax away until it was level with the wood. That's what the instructions advised. This was not like scraping a piece of toast. The surface was unpredictable. The knife would travel easily one moment, meet resistance the next, so that an even hand involved the quick acquisition of an anticipatory sense: when to press, when to yield. One side. Flip the frame. Another side. Next.

As I got better, I slowed down. If there's one thing the gods never tire of teaching me, it's that the moment I bask complacently, I'm likely to get whacked. Two frames done. I clipped them into the extractor. My daughter stood on the step stool to admire my work. Two more done. Two more in. Last two done. I clipped the final two into my spiffy six-hundred-dollar garbage can. The six frames radiated out from the center pole like the petals of a flower. It was nearly seven. Our landlady and her squabbling daughters were usually up by seven, so I thought it a modest courtesy to wait until they stirred. I fed Cheerios to my daughter and Athena—the child added sugar and milk, the cat took hers straight—and made myself a cup of tea. As soon as we heard the daily shrieking and accusations commence, I slid the half moon–shaped silver lids in place.

I set the dial at its lowest position and turned on the extractor. It began to shake violently from side to side, rocking the table as it rattled. I had expected that it would move, but not quite so much. The rattle added a rumble. Was this right? I reached my arms out and held the heaving machine in a tight embrace. What next? Black smoke? Embers? A lava flow? I tried to steady the threatening Vesuvius. My daughter screamed with laughter. "It's not funny!" I retorted. I was wrong.

"It's like a train going through the kitchen!" she shouted.

I heard the crisp sound of metal hitting metal and a thud from inside the tank. I let go and turned the machine off, shaking my head as my daughter cackled. The clips had flown off and flung themselves against the frames. The frames had been flung flat against their wire encasements. I composed myself and tried again. Same results. The hell with it, I thought, deciding that the clips were useless and centrifugal force was responsible for relocating the frames. What if I simply let them stay flat where they'd been flung? The honey would still spin out.

I turned the machine on a third time. It shook so hard that the lavender growing on the balcony at the far side of the house jiggled in time. I dared not adjust the dial upward to increase the speed. If this was low, what on earth was medium?

The extractor pitched. My daughter howled. I held on. Then, another metal clang. The shaking had loosed the motor. Unscrewed it. Unbolted it. All of a sudden, it detached and hurled itself away from the extractor and up into the air. I leapt up and caught it in flight. The motor was now in my hands, but the extractor didn't seem to have noticed. I shut the machine off and stood back. Eventually, the spinning came to a halt. I opened the top of the extractor as if something might jump out at me. There was no monster inside, but the bolts that had held the motor to the extractor were embedded deep in the honeycomb like musket balls in the wall of a Revolutionary War–era tavern.

I sent my daughter off to dress for school and consulted the instruction sheet. There was a telephone number. Aha! With no thought to the hour, I dialed the company that had built this lunatic machine. A recording advised me that this number was no longer in service. This made perfect sense.

The Bee Master wouldn't be in until ten. I glowered at the object that now sat serene as Buddha under the Bodhi tree. What I would do the minute the hour hand was on the ten and the minute hand on the twelve, was to firmly demand that the Bee Master remove this expensive, nasty, and now sticky, abomination from my sight and replace it with a genteel hand-crankable extractor so plain it would meet the approval of the sternest Amish elder. I was done with the world of motors and electricity!

At ten I phone the Bee Master, and after the usual niceties, I made my announcement. "This extractor is a piece of you-know-what!"

"Calm down," said he in that superior, buttery tone men reserve for hysterical women. "And tell me what happened." I did. In a breathless outburst. "It's a good machine," he promised. "Best of its kind or I wouldn't carry it." He nobly suggested that the motor pitching itself toward the ceiling was actually his fault, but that he expected I should always know that any time you used a piece of machinery it was wise to check and tighten up the bolts. If I had a dollar for each of the things I should have always known but didn't, I thought. He asked if I could bring it down to the shop.

"How am I going to unclamp a fifty-pound piece of machinery with an inch of sticky honey in the bottom and no secure top, bring it down a flight of stairs and out three doors, and then load it into my Honda?" He saw my point. "Maybe you can tell me exactly what to do and I'll fix it myself?" The Bee Master knew a bad idea when he heard one. He'd have to make a house call.

Half an hour later he was at my door. He had come with new screws and new washers in his pocket and he secured the motor in place. Still, he was worried about my extractor's natural exuberance. My kitchen table was no home for an extractor. He crossed his arms and his eyes fixed on the twenty-eight inches of mustard-colored Formica that made up my total counter space. He lifted the extractor on its wooden base and set it down, clamping it on the front. There was nowhere for the clamps to attach at the back where a ledge and a loaded spice rack stood in the way, so he jimmied a one-inch-square column of wood under the cabinets, which, for reasons to be found in math classes I never took, made the whole thing stable. But it didn't solve the problem of my extractor's Nijinsky complex. I told the Bee Master how it had leapt on low speed and how, therefore, I was in dread of moving the dial any higher.

"You didn't accidentally put it on full?"

"I swear!"

He examined one of the dials. "They oughta take that off."

"What off?"

He showed me the controls. There was Full and then there was FULL. On the power dial were three positions marked in capital letters: VAR, OFF, and FULL. OFF was in the middle and meant "off." I had assumed that this FULL, as opposed to full speed on the other dial, had meant something like Damn the Torpedoes Full Speed Ahead, or simply, "on." But no. VAR, which I had taken to be a mystery, meant "on." FULL meant "don't touch this at any time or everything you have already instructed the machine to do will be savagely overridden." If you put the speed on low and the FULL dial on, the machine was at maximum power from the bottom of its little spout to the tippy-top of its uppermost bolt, full stop. Why hadn't I thought of that? He adjusted the frames and reattached the clips and switched the extractor to VAR. It purred like a well-fed lioness on the veldt. A lesson in extractorese. VAR made life rosy.

It was midday by the time the Bee Master finished. I offered him lunch, but he had to get back. His wife was minding the shop. I opened the refrigerator and stared at the contents, searching for inspiration. I pulled out some tomatoes and some fresh mozzarella. I needed a bowl and a fork. Impossible. My happy extractor had been artfully secured so that I had no access to either my only silverware drawer or my dish cabinet. And I hadn't noticed. Our breakfast dishes and silver had been washed and put away. The dish drain was empty. We had nothing to eat with or on.

That evening, my daughter and I consumed our first in a long string of terrible meals from the local Chinese take-out on paper plates with plastic forks.

I had blown out two frames in the morning's debacle, but I still had eighteen good ones. Six frames per load. Three loads. So I uncapped and I spun and I spun and I uncapped every moment I was home and awake. I wanted every drop that could or would be mine. When the honey began to accumulate in the bottom, I took a five-gallon bucket and placed it under the spout. I rinsed some clean cheesecloth to strain away the traces of wax and pollen and rubber-banded the cheesecloth over the bucket. When I wanted to take a shower, but did not want to stop extracting, I had my daughter sit in a chair and watch the honey pour out of the spout. It was a boring task, watching the honey drain, but something deep and primitive within me suspected that if this extractor were left unguarded while in VAR, it would slap on war paint and take up its old dancing ways with a vengeance.

Even the emptiest frame still has traces of honey. There were still bees amongst the frames that waited for my attentions. These bees hadn't much of a future. I felt the need to be a decent hostess until I figured out what I was supposed to do with them. At least let these bees have something to sup on, I thought. So I returned the spun-out frames to the honey supers on the balcony. That night, as the bees rested, I did not. My landlady's middle daughter had taken a lover. At night, she crept out her bedroom window to meet him on the lawn. Her bedroom was just beneath mine. I heard the window open. I heard her thump as she hit the ground. I heard the muffled giggles and the whispering and was happy to hear no more until she clambered back inside some time later and there was the click of sash against sill.

Though it was Saturday, I had no hopes of making up for rest *perdu* in the morning. Wondering if my landlady had even the slightest clue about her daughter's nocturnal trysts, I rose with the sun to uncap the last six frames, smoothly uncovering the gleaming treasure buried beneath the wax. I soon forgot about the girl. The getting of honey, like the keeping of bees, asks full concentration. I cleared off the stove, which had been covered with the super containing the last frames, made myself tea in a paper cup and my daughter scrambled eggs, which I served on the now customary paper plate. Then I began again. My daughter hustled into her red jersey and shorts.

Being the only girl on the Under-Eight's travel soccer team meant a lot to her. It didn't matter to her that all the other girls shied away. She played, and the boys respected her for it. Team pictures were at ten. Practice was at noon. Then there was a home game. After the game, we would finish spinning honey. But the day would hardly be done.

I closed the kitten in the bathroom so her tail would not graze the honey, nor her pink nose dip into it as it slowly filtered through the cheesecloth into the bucket. Off we went on errand number one. We returned, expecting an hour of peace. I released the kitten and turned on the extractor, noticing that bees had come around to one of the kitchen windows. Summoned by the intoxicating air within, they hovered outside the screen.

Then came the proverbial knock at the door. On the other side stood an irate landlady with veins afire in her pink cheeks. "You might at least have told me what you were planning," she warbled in an unsteady voice. I supposed she did not mean soccer or apple-picking. "Your bees are mating in the corner of my deck!" My thoughts returned to the resident nymphet. The bees might be up to all kinds of mischief but the one thing I could be sure they were not up to was sex.

I gestured for her to come in. Together we walked to the balcony. It was alive with bees, not just my bees, but bees who had happened upon the spun-out combs and the bees those bees had told. They were commuting down the folds of the ballooning yellow plastic, which was now very loosely covering the remaining super; lines of bees, pockets of bees, all on their way to the sweetness.

If I was alarmed, she'd be more alarmed. I adopted a scientific tone. "Well, there's no queen present, so mating would be impossible."

"Well, they're in a ball. Right under there." She pointed fiercely at the honey super.

"Don't worry. I'll clear it up," I said, as if we were looking at a spilled pot of geraniums. "It's really no problem at all." It was actually a war. Hundreds and hundreds of bees, an uncountable amount, all in motion. The raiders had discovered that the way to the honey was not only through the yellow plastic. Under the balcony and up through the bottom of the honey super was a straight shot. My girls were defending their ravaged splinter nation with all they had.

I distracted the landlady with a description of how the bees use infrared vision to see the nectar. I demonstrated the workings of the extractor, and offered her a seductive spoonful of the goods. She closed her eyes, savoring the blend of sunshine and flowers that was dissolving on her tongue. After ecstasy, one is calm. And so she was.

I dropped my daughter at the field in time for pre-game practice, and promised to be no more than five minutes late to the game. My beesuit sat wet at the bottom of the washing machine. So I put on jeans, a turtleneck, a shirt that buttoned at the wrists and neck, my bee hat, my wellies, and my gloves. Now all I had to do was figure out how to solve this alleged non–problem.

I stepped onto the balcony and was immediately surrounded: over my head, between the slats under my feet, an undulating brown cloud of bees. The war had accelerated. I seized the yellow plastic first. There was no getting at anything else until that was gone. I gave it several hard, rhythmic shakes, wanting to free the bees that would be freed, then I brusquely balled the plastic as small as it would go and shoved it into a black 2.5 mil garbage bag. The loud cloud swirled around my face. Bees covered the netting over my eyes, making it hard to see what I was doing. Opening another bag, I grasped a blown-out frame, whisked the combatants with my bee brush and then, with my gloved hand, ripped the wax right out and cast it in.

The bees were none too happy. I did not want to destroy any frames I could reuse. What to do? My mouth was shut tight. I was sweating. Sweat provokes bees to sting. I brushed off the good frames as best I could. These went into another bag, one, then two, three, four. I could not rid them of all the bees. I tied the bag with a knot at the top and set it on my red rocking chair. Now that it was sealed, the girls inside would feast until they died. Those outside the bag would be unable to join in. The super was finally empty of frames, but still the bees remained. I lifted it and carried it to the roof of the garage that adjoined my balcony. There was no flying in under that. I shoved big black bag number one and all its contents into big black bag number two. Whoever wasn't gone by now was a goner. I tied a knot in the top and heaved the bag over the railing onto the grass, a story below. I scampered downstairs in full regalia, rubber-banded the knot, stuck it in a garbage can, jammed down the lid, ran upstairs, tore off my superhero clothes, jumped into my soccer mom disguise, tossed the kitten back in the bathroom with a little organic catnip for company and made it to the game unstung as the first quarter commenced. I waved. My daughter waved back. I cheered

143

from the sidelines with the other parents. None of them could have guessed what I'd been up to. The whole Herculean deed had taken less than twenty minutes.

We returned to the scene in all ways triumphant. Except for some corpses and a few hangers-on, the balcony was clear. Amen. I heard Athena giving deeper meaning to the word caterwaul. I opened the bathroom door. She bolted. The mirror that had rested atop the back of the toilet was now face down on the floor. There was blood everywhere. The bathtub. The sink. Smeared across the tiles. Fearing she'd been cut by shattered glass, I gently lifted the mirror. Intact. Not so much as a chip. The toilet paper roll had been shredded into fragments as small as moth wings. I stepped out into the hall and found one stunned, humil-iated, striped calico tabby kitten nursing her tail. The fur had been shaved off the tip. The mirror must have landed there. On this day, of all days, catnip had turned her into a tigress. I mopped up the blood. My daughter tended to Athena's pride.

I switched on the extractor with a sense of relief. There was only the last of the last to do. My glistening treasure sunk slowly through the dampened cheesecloth. The pattering sound of honey hitting the metal inside had ceased. I listened hard to be sure. There was no more to be had. Extractor off. I closed the spout. I closed the capping tank. I washed the double-edged knife and wrapped the supers filled with empty comb. They would be ready for next spring. I took two clean plastic spoons, one for my daughter and one for myself, and filled them.

The circumference of the earth is 24,000 miles. To make a pound of honey, bees must gather four pounds of nectar from the flowering world. In a single flight, a bee can visit hundreds of flowers before returning to her hive. Figuring an average flight of a mile and a half in pursuit of the magical nectar, she and her sisters will have traveled eighty thousand miles, three times around the earth and more, to make that single pound. We clicked our spoons together. A toast. To our first ambrosial crop. With unexpected tears in my eyes, I put the lid on our weighty bucket. There were miles to go yet that day, on wheels, not wings. No time to stop at the stopping point. We had a date with Grandma and Grandpa to go apple-picking at their favorite orchard two hours north. Instead of going to the diner for a sandwich, we celebrated at the best place in town. We ate two full orders of prime rib with all the trimmings. And then we hit the road.

7

Fall

There's a whisper down in the field
where the year has shot her yield,
And the ricks stand grey to the sun,
Singing: 'Over then, come over, for the
bee has quit the clover . . .'

—*Rudyard Kipling*

N O SOONER HAD THE MAPLES TURNED THAN they were slapped down by rain. Wet leaves slicked the roads. Children slid their way to school. Instead of reds and oranges and yellows, wintery mauve branches spread like lace over the hillsides, brightly stained here and there like a tablecloth after a meal served with wine. In any other year, my daughter and I might have lamented the dashing of autumn expectations, but this year, there was nothing to rue. Let the gold drop from the trees and vanish under foot. We had our own gold made from bees we counted amongst our intimate acquaintances. And we were nothing if not pleased.

The harvest was over. Our work was far from done. It would, however, be different. Now that we called ourselves beekeepers, fall was preparation for winter; winter, the first step toward spring.

The bucket holding most of our crop remained sealed on the kitchen floor. Letting air in might invite moisture that changed the composition of the honey, so it would have to stay that way until we figured out where to put it. My daughter and I agreed that red-capped plastic honey-bears were too dreadfully common for our magnificent crop. Perfect honey needed the perfect container. My daughter had already designed the perfect label and we had the perfect name: Bee Leaf Honey. Pun intended. It wasn't that there were no jars. You could buy Kerrs or Balls at two places in town and you can't buy underwear at any. Neither Kerrs nor Balls were right. I reviewed the possibilities with dismay until I came upon a hand-illustrated ad about two and a half inches square in the back of an old *Bee Culture* magazine. It promoted a jar called a Wire Bale that came in two sizes. It alone, of all the jars, was worthy. To get it, I had to dial Dorothy in Kansas. Her first reasonable question involved knowing how much honey we had. Who knew? It was in a closed five-gallon bucket. The bucket wasn't full. Nor was it empty. I generously guesstimated four and a half gallons. With this knowledge, Dorothy set to calculating. Gallons to pounds, pounds to ounces, ounces to milliliters. A thunderstorm commenced. The phone went dead twice. But Dorothy went on, figuring the numbers. I learned that there were about twelve pounds of honey to the gallon, 960,000 miles on the wing and millions of flowers kissed. I learned honey weighs about half-again as much as water. A four-ounce jar therefore held six ounces of honey, an eight-ounce jar, twelve. I stared out my window through a curtain of rain. A black Town Car pulled over and Farmer Tom stepped out. He was heading for my house. I begged Dorothy's indulgence as I opened the door. His hair was wet and water ran down in spaghetti strands. He was wearing a charcoal-gray pinstripe suit and soaked brown wingtip shoes. "I just happened to be down the street and I thought I'd pop in," he said. My face must have betrayed me. He explained. "You'll probably never see me again in a suit. I thought you might like to have a look."

I looked. "You're stunning," I answered, grimacing and pointing theatrically to the phone in my hand. I watched him retreat under a blue-and-white-striped umbrella. It was of no use. He was too wet for it to matter and despite the pretense at shelter, he knew it. He didn't bother to avoid the puddles.

The traffic in and out of the hives had died down considerably. Even at the sunniest part of the day, the bees were clustered within, keeping warm. Through the porthole in Har's upper super, I saw a drone standing alone on a frame. He was dying, but not ready to succumb. But other drones seemed in the peak of health despite the lateness of the year. I knew that the workers stopped feeding their idle brothers when the family fortune dwindled. Practically useless in the summer months, drones were sheer liability in the winter. Take take take wouldn't do in times of need. Their presence surprised me. I allowed myself to fancy an exception to the rules. Maybe the boys were dumb lucky. But they weren't lucky, I was dumb. I didn't know enough to recognize a warning.

We saw Pumpkin Pete every morning, five days a week. He wore a silver star straight out of the Wild West and a Day-Glo orange hunting vest. He greeted every child by name, stopped traffic stylishly with a raised palm, and saw them safely across the street. Every day, we chatted. About the land he'd wanted to farm and Farmer Tom had got, about the bees and how they'd do well around pumpkins, about his busy schedule. When Pumpkin Pete wasn't a crossing guard, he was everything else. He ran the town dump, mowed rich peoples' lawns, and fixed whatever needed fixing at a good price for whomever was willing to wait. I'd been on the list for months. A wobbly chair needed gluing, a bulletin board needed screwing into the wall. I was still waiting. He didn't mind being reminded. If you reminded him nicely enough often enough, the job might actually get done.

October was the high point of Pumpkin Pete's year. It was pump-kin time. At pumpkin time, the world was his. He hitched a wood-paneled trailer to the back of his truck and spread the bed with hay. On the bed, he arranged his display. At the center there was always the largest pumpkin imaginable, a pumpkin so enormous that were it hol-lowed out, a small child could play inside. Around the great pumpkin were the kind you might find at a farmstand and take home to carve into a jack-o'-lantern. Even the heftiest of these looked puny next to one of the pumpkins that assured Pumpkin Pete's local fame. Every time a child would remark, "Your pumpkin had babies!" Pumpkin Pete would rock with appreciative laughter and then, forcing himself to catch his breath, reel off the names of all the other kids who'd said the exact same thing and guffaw some more, as if each identical utter-ance was every bit as delightful as the one before, another matched pearl added to a perfect strand. Tucked in around the babies, filling the odd spaces, were a motley assortment of gourds in idiosyncratic shapes and colors that might have seemed more likely in a seven-teenth-century Florentine or Flemish painting. Everywhere Pumpkin Pete went, so went his pumpkins. But even where Pumpkin Pete was not, his pumpkins were. Especially the giant ones. You saw one testing the sturdiness of the pergola at the garden store. You saw one dressed up for Halloween in the window of the local diner. You saw one at either end of Main Street. Everyone knew these were Pete's pumpkins and no one else's and you couldn't help but admire the man's miraculous way with a vine. I'd asked him one day if I could take a look at his pumpkin patch which I imagined to be the eighth wonder, and he replied, "Too muddy in there. I had trouble gettin' my tractor through yesterday." I could tell mud wasn't the problem. There are ways around mud. He was guarding his mystery.

As soon as the beehives were winterized, Red was leaving his wife and children at home and going afar to shoot things. The local things to shoot, which were deer, were dull quarry for Red. Let the locals fell deer. He was after moose. My daughter and I had hunted moose. Determined to see the mythical creature advertised on postcards and traffic warnings, we drove the backroads and highways of Maine, scouring the forests, always a step ahead or a step behind. The only live moose we ever saw was viewed from an untrustworthy distance. It, if it was a moose and not an elk, was at the far end of a large enclosure in a tragic-looking petting zoo along a commercial strip between a shoe outlet and a lobster barn. Red promised us a hunk of moose meat, and I thanked him with pleasure, wholly confident that the elusive moose would easily escape his fire and I would easily escape the moose.

Within the week, I had a happy visit from my friend the moose killer. Bearing moose meat. Just for me. It was frozen in a packet, so I could stall for a while, but not forever.

My daughter had an enlightened response: "Eeeewwwwww!"

I had the same response, but being the resident parent, there were parental things to say about how kind Red had been to us and how helpful about the bees and how the very least I thought we could be was gracious. "We won't have to eat much. We'll just taste it and say that it's very good. Or intriguing."

"What if it's not?" she demanded.

"Intriguing covers everything," I explained. "And it's not as unkind as 'interesting'."

"Or 'different'," drawled my seven-year-old. She'd heard it at school. She'd figured it out. Different. The thing that is not the same as all the other things. The thing that stands out. Different, in this town, is not merely descriptive or value-neutral. "Different" is what they say around here when what they mean is bad.

I set about searching for venison recipes with sweet, complicated sauces guaranteed to mask the awful truth. At bedtime the night before our agreed-upon moose night, I took the hospital-green paper packet out of the freezer. The next afternoon, there was time before soccer practice to arrange the wine and dried fruits I had purchased for the sauce, but not to cook them. The ponytailed athlete returned, hungry enough to be almost willing. I opened the paper packet to inspect that which was about to be roasted and drowned. This was not what I had expected. Not at all. This couldn't be roasted. This couldn't be drowned. It was ground moose. Mooseburgermeat. My daughter hated burgers of any kind. There was only one remaining option. Mooseloaf. "Who knows?" I offered weakly, "We might love it." My daughter glared. I decreed that three bites apiece was an honorable amount, respectful of Red's generosity and enough of a sample from which to determine whether or not mooseloaf should become a household staple. With the aid of a certain tomato product manufactured by Heinz, we were able to reach a unanimous opinion. Mooseloaf was decidedly . . . different. All meanings applied.

Where there had been flecks of autumn bright there was now a smudge of faded curry. You could see the ground cover on the mountain. The view was wider and emptier. The river was moss green and so still Narcissus might have knelt at the banks to admire his beauty without the distress of an errant ripple. There were ducks bottom-up in the inlet, flocks of birds in the field and in the yard. There were Canada geese everywhere, some moving on, some moving in. Har was so quiet that I feared there were no bees in the hive. I moved the entrance reducer, a wooden block with a notched exit the width of two workers that is wedged between the bottom board and the entrance so that mice stay out and warmth stays in. I saw no guard bees. I peeped in the portholes. No one up. No one down. Now troubled, I opened the hive cover and saw a few bees hiking across the yellow insulation I'd installed for warmth. There were a few in the upper honey combs, very few. I hoped they were clustered, warming each other, in the bottom hive body.

Har, being my strongest, most populous hive, was also the one that might swarm. If Har wasn't big enough for the girls to live happily, the scouts would look around for a new place to live while the workers made a new queen. The old queen would choose a warmish day and gamely lead half the hive to *terra nova*, leaving the anointed daughter to take over the castle, to mate, and to continue her good work. But if Har had swarmed, there would be a poor selection of drones, if any at all, to enrich the new queen's eggs. She might remain a virgin. If so, the hive would face dysfunction first, then death. And the old queen with her minions would probably starve so late in the year. With what would they fill the new white comb they'd construct in a space between barn walls or the hollow of a hospitable tree? Pollen would be scarce. And nectar? They'd die too. It was a sad thing to contemplate. Jafenhar and Thridi seemed stronger, but I wasn't sure I could tell a winter mood from loss.

Thunderstorms threatened. My spirit suffered. I have known these four seasons for many years and still greet each one like an unexpected guest. Winter is the one I love the least, for it is too long to live with no green but that of the muted pines. The once-joyful Persephone had returned to her chilly husband, Demeter to her lonely grieving. I still had my daughter at my side, but my bees were out-of-doors with only themselves to rely on. No matter that for millions of years they had needed nothing more. I felt a helpless maternal ache. Poor bees, I thought, understanding how certain sorrows echo through all time.

"Did you see the dawn?" my daughter asked. We were up early. It was an important day. Her team was undefeated. Now they'd face the best players in the region to determine the U-8 champions. I had seen the chiffon pinks wavering pale to bright, the rosy orange, then pale bittersweet. Color had retreated from the earth but not the heavens. In an hour the sky was pewter, weeping tiny pellets of ice. In spite of the weather, I was glad the tournament went on. The team charged through rain and even snow, sleeves pulled over their little fingers for warmth. And they won again. For the first time in town history, the youngest local team took home the big brass trophy.

The local newspaper headline crowed: U-8 BOYS SOCCER SQUAD REGIONAL CHAMPS, ignoring my proud daughter's presence completely. My daughter's hair falls free to her waist. She cannot be mistaken for a boy and she knows it. She did not weep. She raged. She'd been at every practice. Played every game. Done her part. How could the newspaper print such an obvious lie?

"Dear Editor," began her letter of protest, "The newspaper said: The U-8 <u>BOYS!!!</u> soccer squad won the championship. In fact, there was a girl on the team and it was me."

With her first published words came a second victory that meant at least as much to her as the first. On Main Street, people who knew her and those who just knew who she was took note, stopping her to remark about her pluck. When it was time for the team victory party, the volunteer coach, who had three growing sons to feed and was far from a wealthy man, surprised the team with engraved trophies he'd paid for on his own. Every boy got a golden boy kicking a golden ball. The coach presented my daughter with a golden girl, golden ponytail wild in the wind kicking with as much determination as any boy.

As will sometimes happen at this odd time of year, the weather reversed itself. Where there had one day been a solid chill, there was now late springtime warmth and gaiety. The bees were merry like everyone else, finding excuses to leave the hive though there couldn't have been much to gather. Five drones sunbathed on Thridi's bottom board. Drones in mid-November reminded me that all time is borrowed time.

Because Red had persuaded me that the traditional wrapping of hives in tarpaper was not a good idea, I made other plans. The little portholes in the hive bodies and the little slot in the entrance reducer would be able to act as windows, letting air in. But they might also act as a come-hither wink to little mice seeking shelter when the weather quit flirting and cracked its icy whip. Seeing how they make quick work of intrusive wasps and bumblebees in the summer, one would think the bees would defend their hives against a furry houseguest, but in the winter they might not because in order to stay alive and warm themselves, they must remain in cluster. They form a ball with the queen in the center. Huddling tightly together, they not only create their own insulation against the cold, they generate a modest amount of heat. They would certainly notice a mouse. They do not hibernate like bears. But for individual bees to break cluster and chase a mouse away in thirty-degree weather might prove to be a fatal decision. Unlike their keepers, bees have never been known to be wasteful or foolish with their lives.

The plan, Red's and mine, was to purchase something called hardware cloth so that the bees could get out on a warm day for what is politely called "a cleansing flight," but the mice could not get in. That involved knowing what hardware cloth was, which I did not. I didn't have the nerve to ask Red. I had just returned the wire and tools he'd lent me to build the electrical bear fence. My buddy at the True Value looked crestfallen. He didn't carry what I wanted. He took the failure personally. "I'd love to help ya," he said. "But we don't have much call for hardware cloth, I'm afraid." I comforted him by discovering something else I really needed, an extra large Pyrex measuring cup, and buying that.

My daughter and I ventured southward. We spotted an uncharted hardware store with the right number of ladder-hatted vans and blue trucks in front of it. I cast myself on the mercy of an unknown hardware guy who sent us out back to a barn filled with wire and lumber. Strange hardware guy number two showed us to the side double doors and pointed through the drizzle at hardware guy number three. He stood next to several giant rolls of what looked like window screening for people who love mosquitoes. "It's to keep mice out of my hives," I explained.

"Well, I'd go for the quarter- or the half-inch, myself. But what you really need is the Pied Piper."

The Pied Piper struck my daughter as the right idea, but I, firmly anchored in banality, settled for a roll of both the quarter- and the half-inch and a stylish pair of yellow-handled wire cutters and a staple gun, raising the price of the expedition from cheap to not-so-cheap.

Wire cutters are a delightful toy for those who find a different pair of scissors necessary for every occasion. I happily snipped six squares each of quarter- and half-inch hardware cloth, proudly adding my wire cutters to my delicate nail scissors and my pinking shears, my paper scissors, my fold-up Japanese mini-scissors, my little Victorian bird-headed thread cutters, and my fancy German gold-and-silver superscissors that were purchased during my few brief weeks of trout fly tying.

I gathered my admirable hardware cloth squares and took them, with my staple gun, out to the field the next morning. Something was different. Something was off. I looked around me, trying to determine what was missing.

It was the table farm. Gone, down to the last sawhorse. Where had the plants gone? Had they at last languished to death?

I spotted Farmer Tom leaping over the spindly remainders of what he had put in the ground, bounding toward me with the dread eagerness of a wet dog. He held out a plastic bag, indicating that I should smell it. Oregano. And some aromatically unbasilesque basil.

"Now that I use this in my pesto," he enthused, "People can't get enough of it!"

I asked if he'd be farming this land next year. It was up in the air, he said. If the barn was fixed up as an apartment, maybe. "A farmer basically has to roll out of bed onto his land because there are always things to check in the middle of the night." A familiar song. "You have to always be there, at the ready!" But then you have to do something, I thought. He'd met yet another rich woman. She lived six months in Paris and six months nearby. She had property that might take well to farming. She was more than a little interested. When she asked how much money he'd need to work her land, he'd answered "Roughly a hundred fifty thousand." And she'd said, "Send me your proposal." I'd seen the seductive radiance of his polished pitch, so I wasn't surprised to hear that the widow of a famous writer also found herself entranced by his possibilities. As did a professional fund-raiser, who could, with a few well-placed telephone calls, make real his passion.

"I'll go where the money is," he confessed.

Ceaselessly attempting to charm rich women into bankrolling his vividly enunciated vision was more gratifying to Farmer Tom than the long hours and dirty, hard work of actual farming. I pictured Jeeves handing him his morning espresso as he rolled out of his bed, into his morning coat, and onto his farm. Once more, he uttered his *cri de coeur*, "Farming is my dream!" It most certainly wasn't his reality.

Having chosen to try out the quarter-inch hardware cloth, I stapled the three-by-three-inch patch over the portholes without once piercing my fingers. The workers could get through, but the little squares were too small for drones. In theory, this was not a problem. But there was a fellow at Har who was going through drone hell and I couldn't bear to watch it. No matter what angle he took, he could not fit his bulk through one-quarter-inch square. Figuring he was not long for this world, I showed him a last kindness, lifting the wire so he could feel the sun on his wings. There was bright orange pollen coming into Jafenhar, the hive that had struggled through a supersedure. Even in November, if the weather was dry and warm, these girls and their new queen were unstoppable. How they must have searched to find something still abloom. And yet they did.

8

The Chill of Winter

"Yes it's a very good bee hive," the knight said in a discontented tone, "one of the best kind. But not a single bee has come near it yet. And the other thing is a mousetrap. I suppose the mice keep the bees out—or the bees keep the mice out, I don't know which."

—*Lewis Carroll, Through the Looking Glass*

B Y MID-DECEMBER, A VISIT TO THE FIELD was like returning to the scene of a past happiness. The place was still there, still beautiful, the joy was still with me, unforgotten, but not to be touched by my hands or seen by my eyes. It lived in a time, not a place. The field was dormant now, a sandy straw color. Rotting squash. Flowers dead on their stalks. Shriveled basil. We had been far away for a few weeks. I felt a terrible loneliness for my bees. The false spring had persisted while we were gone. There were buds on the forsythia bush outside my daughter's school. But the buds would freeze that night.

My return to the hives was not entirely comforting. Thridi was quiet. I saw a dead bee in the top porthole and four dead bees on the bottom board. I swept them away and adjusted the entrance reducer, assuring myself that death happened year-round. It was just more noticeable now, when so much else was dead or dying too. Har was more troubling still. The hive was very quiet. That was to be expected on a chilly day. But why had the entrance reducer fallen to the ground? I knelt and looked along the bottom board into the hive, a trick I would never have attempted with guard bees actively patrolling the entrance. I saw dead leaves and debris from the outside. The wind might have blown it in. A mouse might have carried it. I took a stick and tapped at the sides of the hive body. Nothing stirred. I scraped the stick along the bottom board inside the hive, clearing away all the junk, but detecting no movement. If there was a mouse, it was out. And it was going to stay out. I shoved the entrance reducer back into the gap, firmly lodging it so that I myself would have to struggle to get it out come spring. Only Jafenhar sent someone to observe me. As I turned the entrance reducer around so that only the smallest hole faced out, two inquisitive sentinels surveyed my work. I left them reluctantly, feeling overcome by both devotion and helplessness. I couldn't stop the cold and hardship from coming when it would.

We spotted Pumpkin Pete near the dairy case at the IGA. Easy to do. He was fitted out in his brilliant orange jacket. I had no doubt he'd know about Farmer Tom's wavering relationship to the farm, and I asked if he'd been thinking of trying his luck again. "I drove over to talk to that guy who runs the Land Trust. I forget his name. He's kind of out in space." Pumpkin Pete's hands were shaking. Broken capillaries mapped the skin on his nose. "We don't see eye to eye exactly. He says they want someone certified organic for horticultural. Has to be certified. They want flowers. So they're gonna advertise in a horticulture magazine."

I suspected the Nature Lover's hand in this. "You're organic, aren't you?"

"Yeah, but you gotta take a test." I could see that the thought of being quizzed on his knowledge was terrifying. "Take a test to get certified. I started out small. I didn't think about it. I didn't think I was going to get so big I'd have to take a test."

I didn't know what to say except I was sorry. It was five days before Christmas. He'd been fixing up bikes left at the dump so that kids who couldn't hope for much would have something good under their trees. But poor Pete hadn't a chance. Not on that land with those people.

Down at the diner, I ran into one of the farmers who'd been farming all his life, one of the old guys who got a kick out of a woman like me being a beekeeper. I wanted to talk to him about helping Pumpkin Pete, but I couldn't. There were ears at every table. I waited until I saw him a second time that day, and waved his truck over to the side of the road. He listened to my tale of poor Pete's dilemma, his head bobbing slowly, asking no questions. "He don't mean no harm to anyone. But he likes to tell stories."

"I feel kind of bad for him."

"Oh, he likes to tell a story," the farmer repeated. I didn't understand. Maybe he drank a little more than he should. His nose told me that. But a story?

"He's had some land." The farmer gave some thought to whether he was going to trust me with the rest. "Pumpkins and gourds. Wasn't organic at first but by the time he got done with 'em they was. Just let 'em grow up 'longside the weeds. Then he realized that wasn't harmin' the pumpkins none. Nature's been organic for a long time. Now he's a real good man, Pete. And he knows his gourds. But those giant pumpkins?" I knew the ones. Everyone knew them. "He don't grow a one of 'em."

Impossible! This could not be true. He was Pumpkin Pete. Famous Pumpkin Pete.

"He knows I know, so he told me. He said, 'I tell everybody I grow 'em 'cause it's better that way.' But he don't."

"Where—" I was too shocked to form a sentence. "How—"

"He buys 'em. Over in East Millersford."

I laughed. The farmer laughed with me. Pete had the whole town believing in his magic. And why not? As far as I was concerned, there was nothing to do but delight in his ingenuity. Some people might think he was doing wrong. But that would only mean they didn't understand. Pumpkin Pete was grand, a visionary in his way: imbuing the everyday with magnificence, feeding dreams of possibility, stoking wonder in children. He knew fate had only given him but so much, but that didn't stop him from giving the town an annual marvel that was worth more in the merriment it sparked than grim submission to his limitations could ever be. Pete was a soulful grower of little pumpkins and gourds who wanted to produce giants. And so he did. Just not quite the way we all assumed.

On Christmas Eve, we hung our stockings by the glass door to the porch because there was no fireplace. Santa didn't seem to mind. He sampled the cookies my daughter had baked and fed the carrots to the reindeer. There were a lot of stockings to fill. In addition to those put out for ourselves and the small assemblage of friends who attended our annual stress- and relative-free Christmas morning breakfast, there was a stocking for Athena the kitten. Hers was filled with multicolored mice, the kind upholstered with eerily realistic, possibly real, fur. She'd had lots of toy mice before Christmas. Playing with predatory violence, she loved to cast her toys up in the air and fiercely attack them where they fell. She tirelessly batted them around the house until she inevitably lost them down some invisible black hole leading to another dimension where they could never again be found.

On Christmas morning, in our house as in most houses, it is the child who rises before the dawn and rushes to the tree. Hearing the expected rustling, I smiled and nestled into my flannel sheets, closing my eyes as I waited for my daughter to announce that she had been well remembered by the man in the red suit. I had misinterpreted the commotion. My daughter was still in her bed. She was cheering Athena, who was crashing about. I listened. I heard a squeak. Scampering.

"Mama! Santa brought Athena a real mouse!"

It was Christmas. She was seven. What could I say? "Such a wonderful present!" A mouse in the house doesn't horrify me anymore. It's the country. Old Ruffy had always taken care of our mouseish guests. Brutally. He liked to eat the head and leave the rest for me.

The festivities had begun. Athena chased her handsome gray Christmas mouse up and down the hall, unflustered by human onlookers. We jumped out of the way as the two heedlessly scrambled over our slippered feet and through the house. Athena clearly failed to understand that nature intended her playful predatory zeal to turn real. When she got close, she patted the mouse timidly with her paw, as if it were a new, intimidating version of her favorite furry toy. When it fled on its little legs, Athena was overjoyed. She went bounding after it, pacing impatiently when it hid. Her Christmas mouse had more moxie than any mouse I have ever seen. It stood on its hind legs and scolded Athena, fighting back with its tiny forepaws.

A guest arrived, taking the madness in stride. We settled in the kitchen while the cat and mousing continued in the bathroom and the hall. Our next guest was another story. She froze at the news of Athena's new playmate. "I. Am. Totally. Phobic."

"The mouse is long gone," I lied.

"It went right back into the wall," lied my daughter.

"I think we scared it off," lied my houseguest.

I poured her some of my warm *Jul glogg*. She sipped, her head tilted cautiously as if she were listening for the telltale pitter-patter, until she regained her normal color. Athena sauntered into the kitchen as if it were just an ordinary day and began nibbling the pellets in her bowl. I excused myself, anxiously inspecting the bedrooms, the back room, the bathroom, the hall. Not a creature was stirring . . . and no sign of a carcass.

The way to end the ordeal of Christmas shopping is to become a beekeeper. My daughter and I gave identical presents to everyone who merited inclusion on our most-favored list: a beautiful hand-labeled old-fashioned glass jar of our own first crop of honey. It was a simple gift, a humble gift. Like pure platinum. Or Tiffany diamonds. Only tastier. Not including labor, mine, my daughter's, or the bees', a single narrow four-inch-high jar cost approximately $200 to produce. It was worth it.

The lights on the Christmas tree became more brilliant as the day outside grew darker. At sunset, we bid our happy phobic guest a merry Christmas and sent her on to her next engagement. After we said our good-byes, we went to sit by our tree. Out from under my filing cabinet came the Christmas mouse, who had most courteously waited for her departure. Athena and her playmate began again. When Athena astonished herself by actually catching the mouse, we began to worry. This mouse had too much character to die an ignominious death. When Athena backed the Christmas mouse into the corner where the shoes and boots were piled, I grabbed a box, clapped it over the mouse and called for a lid, which I flattened and slid under the box. I hustled the captive out onto the lawn and set it down on the grass next to a squirrel's cast-off ear of corn. The brave mouse shook, from fear, the cold, or both. I was no longer certain I'd done right. It wasn't far from the house, I told myself. The mouse could find its way in if it needed the warmth. I hoped it did not find its way back to the cat.

Athena nosed around the house looking for her playmate. Finally she gave up, settling for a blue, legless facsimile with nobby eyes and a detachable tail. After a few minutes of vigorous tumbling, she gave up. This mouse was not that mouse.

But neither was it the mouse we had yet to meet.

The river was white and gray and the trees looked like crusted wires. The crows laughed, walking across the field like elegant women in black waiting to toast the new year a few hours later, at midnight. They were the only evidence of life in the color-leached surroundings. Partway across the field, I saw mansized footprints in the snow. I walked next to them, wondering if someone had gone out to look at my bees. But the footprints stopped at a certain point and went where I wasn't going. Here, our footprints diverged so that, if another walker followed our steps, he would conclude two people walked together until they went their separate ways. The possible misperception pleased me.

Jafenhar was quiet. It showed no signs of death. I pictured the new queen reigning with winter restraint and a firm hand. She had become my favorite bee. Har, my boisterous Har, was in trouble. Someone, and

it wasn't hard to guess who this someone might be, had chewed a round hole all the way through the entrance reducer I'd lodged to block the access of wind and unwanted invaders. I could see the chisel work of tiny teeth. This creature must have been desperate. How long had it worked to get out of the wicked cold into the warmth of a beehive?

I was not overcome by compassion. I knew mice could do a lot of damage to a hive, gnawing the wood, chewing away comb to make space for themselves, even eating precious winter honey.

Red told me that he always stuck this stuff called "One Bite" right near the entrance to his hives and in his five years of beekeeping he'd never had a mouse problem. I brought my morsel of knowledge to my buddy at the True Value.

"I'd be real careful with that," he said. Then he reconsidered his cautious statement and made it more cautious still. "I'm not sure I'd recommend that to you."

I was intrigued. I could not recall a time when I had walked into a store and been told not to buy one of their products. I read the package. This stuff wasn't a little bit of something to get rid of an out-of-place mouse. This stuff was *poison.* It could kill just about any living thing that came in contact with it. The field, which might or might not have been pesticide-free before Farmer Tom neglected it, would be sullied. The cats, squirrels, chipmunks, dogs, foxes, raccoon, deer, coyotes, opossum, geese, turkey, ducks, crows, not to mention miscellaneous critters that might happen by in search of food could be done in as well. What if a bee taking a cleansing flight should rest six tiny legs, however briefly, on the One Bite? Into the hive the poison would go. Into the bees. So in order to kill the mouse, I'd kill the hive? Furthermore, the package warned that the poison should only be employed above the high water mark. It could also poison the river. Since my hives were two hops from the river and undoubtedly in the flood plain, even if I'd been stone stupid and had a heart as cold as a serial killer, the directions advised against.

It was twelve degrees Fahrenheit. We'd had high winds, ice storms. A very short time ago, if you calculate time by how long bees have been on the earth, a human mother and her child found shelter in the home of other creatures around this time of year. The mouse was staying. Let it be warm. I'd deal with the damage come springtime.

We were all sharing the cold. Pumpkin Pete did not desert his crossing-guard post on account of a little weather. He wore a fur cossack's hat with the earflaps down. Over his blaze orange vest, he wore a blaze orange coat. Under his windproof fleece gloves he wore Thinsulate gloves. Even through two pairs of gloves, his hands still shook.

The local paper featured a photograph of four mallard ducks who had moved onto Main Street: two jaunty males wearing their green iridescent caps and two modestly attired Mamie Eisenhower females. The Sunday after the article came out, there were three. One of the females had vanished. The rest were hanging out in front of the bookstore greeting the local intelligentsia. The whole town worried when the duck disappeared. Had it become someone's roasted duck *à l'orange*?

There were inches of ice everywhere. Despite a Dead Sea's worth of salt on every flat outdoor surface, the only thing more hazardous than driving was attempting to walk. School absolutely had to close.

This struck me as the perfect time to visit the bees. Reasoning, and not reason, accompanied this decision. The Bee Master, like all true agricultural men, was on intimate terms with the weather. It told him things it did not tell the likes of me or the National Weather Service. What it told him was that despite the frigid evidence before us, we were on the verge of a January thaw. If there was going to be a January thaw, there was a chance the temperature would break fifty degrees Fahrenheit and my girls would want to refresh themselves in the warm afternoon air. They urgently required roomier hardware cloth through which to slip or they wouldn't slip at all, judging from my last visit to Thridi.

Not expecting apian companionship, I took only my dreadful hat and the netting for cover. My coat, my hat, my winter gloves would keep me warm. There were bird prints, tiny arrows with elongated tips, which got denser near clumps of grass. I wondered if they had huddled there. Along my path, I saw the glazed paw prints of creatures obliged to leave whatever warm place they had found in search of a meal, but no print as big as a man's or as small as a mouse's. It was five minutes' work to tear off the old hardware cloth and staple the new. I nosed around Har, looking for signs of what the mouse was up to, but there was nothing to see. At Jafenhar, I saw a lone worker venture out to dispose of a dead sister. She took the bee down to the snow, but did not just drop it there. She took some time with it, adjusting the way it lay before she returned inside. Perhaps she was lingering in the light— the day had warmed, the wind vanished—perhaps she was tending to the dead. This healthy bee, though she did the work of an undertaker, was a sign of a hive well and alive. I took off my hat and netting and pressed my ear against the cold wooden supers. I heard humming within. Joy! I could have waltzed to that vital monotone.

Moving on to Thridi, I found silence. I lay my ear against the wood once more. Silence. I checked the entrance reducer. No gnawing.

Probably no mouse. I rapped on the super, trying to raise a response. Silence. And then I got it in my mind to lift the outer cover and check the insulation. An excuse to peek. Knowing it was an excuse did not stop me. I promised myself I'd be quick about it and casually tossed my hat back on my head, not bothering to fuss with the netting. The bees were outraged. Flew up gaps in my sleeves I hadn't known were there. Stung me once, twice, three times, four. "Here I was trying to help!" I huffed to myself like a wounded martyr full knowing my carelessness was the true cause of my pain. "Fine. Freeze," I snarled indignantly. I couldn't remove the stingers in my arms, up my sleeves, without undressing in the snow. A bee flew under the netting. She was buzzing near my cheek. I tore off my hat. Two more bees burrowed in my hair. Holding my breath, I secured the inner cover. I listened to the buzzing on my scalp. I combed my fingers through my hair. I could not find the bees. I could close my mouth, but what about my eyes, my ears? This was a kind of fear I had not known before. I stumbled away from the hives lest these call others to their aid. I bent at the waist, and resumed the blind search, raking my hands through my curls. Nothing flew out. My hair is so thick that one summer a fallen baby bird, frightened and disoriented, flew into it, mistaking the abundance for a nest. A bee could be there for days. I hung from my waist until I was flushed and woozy. Perhaps the car mirror would help. I crossed the field, taking care that my steps were deliberate and well placed. I did not want to compound my problems by falling on the ice. My hair stood straight away from my head in long spikes. I looked like I had put my tongue in an electrical socket. I bent and combed some more. After a while, my head was quiet and though I had seen no one leave the scene, I dared to hope the bees had found their

way out. Rattled and in pain, I heaved myself into the driver's seat. It was all my fault. I was a world-class idiot. All I wanted to do was go home, remove the stingers, shower, and repent with a shot of Lagavulin, my favorite single malt whiskey. But it wasn't going to be that easy.

I had assumed that the tracks I'd made on the way in would be the tracks I'd follow on the way out. I was wrong. There were bigger tire tracks over by the inlet. Perhaps if I backed into those and got some traction? My car began to skate dangerously. I shifted to first and steered for the snow, my other choice being the water. I made some calculations and angle by angle, turned the car around. At least it now faced the road. I started. The Honda stopped. I was fifty feet from freedom. The sky was turning pink. I tried again. And again. I walked up the hill to the road and began waving my arms. People waved back. I added desperate jumping up and down. Finally someone realized I was not just a friendly moron and pulled over to help. It was a dad with his boy in the car. The dad tried three times and failed three times. He drove me to my house. I thanked him for his trouble and called AAA. They refused to help. "We won't tow unless the drive is cleared and sanded." Citing the illogic of agreeing to tow only where towing is unnecessary and the hundreds of dollars I'd paid without asking for more than a TripTik, I demanded assistance. No sooner had I put down the phone than the lurking bee in my bonnet materialized once more. It hummed as it wandered the thicket on my head. Whistling in the dark. I was in no mood for nonsense. I shook my head fiercely. It fell onto the carpet. Athena pounced. I tossed her prey into the toilet and flushed.

Six-thirty that evening the tow truck driver appeared at my house. By 6:35 I was behind the wheel of my car as it was being dragged up a hill in the dark by a single metal cord. If the cord broke, I was in the icy water confronting my mortality. The cord did not break.

I did. Within three days, my neck was so swollen I looked like Benjamin Franklin with gout. And that was just starters. I called the doctor. It sounded "viral" to him. He suggested I use propolis, which I could purchase in paste form for $8.50 a jar.

Having scraped it off the hives on many occasions in order to open what my hardworking girls had deliberately fastened shut, I had lots of propolis. I proposed using my own, confessing my only hesitation. If you put propolis, known as bee glue, in your mouth, it sticks to your teeth.

"That's probably the wax," said the doctor. "If you put it in vodka, the propolis and wax will separate. Then just pour off the vodka."

"Why bother?" I replied.

The prospect of employing a medically useful product of my bees' labors to cure my Ben Franklin–itis combined with the prospect of using up a bottle of vodka left over from a party three years before immediately improved my outlook. I dumped my Baggie full of propolis into a mason jar and pulled the heavily frosted bottle of Absolut from the freezer. No sooner had I poured the vodka over the bee stuff when the wax began to spin away from the solid amber propolis, floating to the top as the center of the jar produced a frosted band of ice. It looked like a liquor parfait. Propolis. Iced vodka. Wax. I left it to sit for hours just in case there was more separating to be done. Infused with health-giving tree resins, the Absolut became pale gold. I scooped off the wax and took a swig. Hot going down. Spooning my now pure propolis out of the bottom, I popped it in my mouth. And that was the end of the do-it-yourselfer optimism. The propolis clung to my teeth and between my

teeth and to the roof of my mouth and the sides of my mouth and dyed my teeth yellow besides. It stuck to my fingers and whatever my fingers touched. I brushed and flossed and brushed and flossed and brushed and flossed and that night, when I got out of the bathtub, I found propolis stuck not only to the bottom of the tub but to the bottom of me.

Just as I was contemplating another nip of propolis-infused vodka, to be combined with vitamin C-infused orange juice, to be combined with escape-infused sleep, Farmer Tom called to tell me he'd been in Pennsylvania as the guest of a millionaire with four hundred acres that she might be amenable to giving over to Farmer Tom's organic fantasy. Farmer Tom would be willing, he told me, if and only if the millionaire would give him "a couple hundred thousand dollars to do it properly" and agree to let him hire a foreman so he could step back and cultivate the vision without wasting his talent by actually sticking his hands in the soil.

The Bee Master's prediction proved correct. In one day the air went from winter to warm. Rain fell for thirty-six hours. The ice that had held the top ten inches of the river in place melted. At once. You could hear it roar and crash in the night. Dreaming of beasts in combat, I woke to ice run aground and deep green-gray wide fast running water. Hidden powers exposed their might. The river tumbled over its banks. Roads became hazardous ponds closed to traffic. I wondered if my hives had been swept away.

As soon as it was sane to do so, I drove to the field, parking at the side of the road to avoid the driveway where inches of mud pudding had replaced the ice. The hives were standing, but they'd been hit. The supers were wet, as were the bottom boards and the entrance reducers.

The cement blocks upon which the hives rested were soaked. The river was no longer up around the hives, but it was nearer than before. The water must have rushed up, embracing the hives and receding in seconds, minutes, or hours. Who could say? I walked the fifteen or so feet to the water's edge. There I saw great broken slabs of ice, easily eight inches thick, the top inch crusted with snow, the rest, sheer, bluish-clear like bubble-flecked glass, preserving a leaf here and there. The force of one of these blocks, five feet across and twelve feet long, would have been formidable. I was lucky a barrier of trees stood between the ice and my soggy beehives. I did not see any new dead bees, but that didn't mean much. The bees were still. The inside of a hive must remain warm and dry. Was there even a chance of that? None of the bees were out taking the air. I hoped they were still alive. The flood made uncertainties more uncertain still. I had put the hives in this spot because it was lovely near the river by the hills and I felt happy there. I had not considered the breadth of a field when carrying the heavy weight of honey supers. I had not considered the ferocity underlying the sparkling calm that in its beauty drew me near. Nature can be harsh as readily as she is kind. If my bees survived the consequences of my aesthetic sensibilities, I would make amends. I would move them where they could thrive without fear of the preventable.

9

A Mulched Footpath

There is no room for death:
alive they fly
To join the stars
and mount aloft to Heaven.

—*Virgil*

FTER WAITING FIVE MONTHS FOR HIM TO come around and do a few repairs, Pumpkin Pete finally showed one snowy school day at ten. He came without his tools, explaining that they were in his truck and his truck was in the shop. I suggested we get down to work. He looked at me skeptically. "Do you have pliers?" Needle-nose and two blunt. "Do you have a wire cutter?" Of course. "Do you have screwdrivers?" In thirty-one flavors. "What about a hammer?"

"What kind do you want?"

"I shoulda known," he said approvingly, "she comes with all parts and accessories." Unable to disguise my pride at not only owning these tools but knowing what they were, I threw open my toolbox as if it contained gold doubloons.

His conversation turned to the land. "If your farmer there had put in pumpkins, I tell ya, it woulda cost him pocket change, the whole field. He'da had to keep the weeds down at the first, then after a while the plants woulda overpowered the weeds and it wouldn'ta grown so bad. I arranged to have five acres this spring coming up. Go there every other year so the land can have a rest. You could keep your bees that place if you wanted. If the flood plain becomes too much . . ."

I might have known. Pumpkin Pete, like the other old farmers, didn't miss a thing. While I was making my aesthetic judgments and situating hives near riverbanks, they were probably cackling louder than a house full of hens. I made sure Pumpkin Pete left with a jar of our extraordinary honey. Word would get around about that, too. Though I might be a fool, I wasn't a fraud. That honey was the real thing.

My daughter came home from school and did her homework. We ate dinner. She had a bath. We read a chapter of *The Horse and His Boy*. The phone rang. It was my mother. My stepfather had suddenly gone ill and pale. He was in the hospital. The next time the telephone rang, he was dead. We left our life here, dishes in the sink, and drove north. He was my daughter's only living grandfather. They were each other's favorite playmates, eagerly conspiring to eat chocolate whenever and wherever they could, even when Grandma firmly said no.

At the funeral, my daughter insisted on peeking into the open coffin. I let her. She studied the waxen face within, silently, fearlessly, noting the makeup, the false coloring, the lipstick, going back for another look and a third. At the age of seven, she had written her second letter to heaven. She placed it beside the body and then took my hand, drawing me away from the mourners for a private conversation. "That's not Grandpa," she confided. "That's an actor." I knew what she meant. She tolerated the eerie falsity of a dandified corpse passing for a man she loved and adored. And then she noticed a photograph. It was his picture propped near, the contrast between a discarded shell and this, a moment taken out of time, a stolen second from before, when he was whole, body and spirit together, a record of the real Grandpa standing against a blue sky, the real Grandpa in an Irish cap smiling at something he saw and we, the viewers, did not, that made her inconsolable. Having lost Ruffy, she well understood what Grandpa's being dead would mean, how *gone* gone was. She also knew how to cherish more than memory. I found myself grateful to our old gray cat for dying first. She had already learned she could say good-bye to someone without losing all she had. As long as she lived, nothing could take away the love. That was hers to keep.

It wasn't until Valentine's Day that I made it out to the field. This was to be my valentine to myself. It was fifty-five degrees. My daughter saw a bee taking a cleansing flight outside of Jafenhar, so we knew that the hive had bees. I lifted the cover off Thridi. But there were no remaining signs of life, nothing except plenty of uneaten honey. I told my daughter to step back and removed a frame toward the center, where the bees would most likely be clustered. A third of the frame was covered with worker bees who looked like they had been flash frozen. I whacked the outside of the super with my hive tool. No sound rose up from within. They were dead. All dead.

I moved to Jafenhar and opened that. Two bees climbed up to investigate the intruder. Good. Queen Mab, my supersedure queen, was still working her magic.

Score: Death, 1. Life, 1.

Har was tricky. It was hard to tell. The bees had neglected the Terramycin patties that had been left for them to nibble, but that might be because they didn't like the taste. The bees had glued the frames to the hive, packing the propolis so well that my efforts to pry the frames free was causing the wood to strain. I didn't want to crack the super. If I was going to make that kind of a mess, it would have to wait until spring. I couldn't see down through the frames to the mouse I knew had gnawed its way in. Nor could I see any bees. The bee picture looked bleak.

My daughter and I walked back to the barns. There was a car parked behind ours. A man had driven two hours to sit behind his steering wheel, charcoal in hand, and sketch this winter landscape. He asked what we were doing. I told him. He was impressed. "I've never met a beekeeper before."

"You maybe still haven't," I mumbled.

I spent the evening studying bee diseases and looking at disgusting photographs of things gone wrong in hives. The Bee Master hypothesized that Har could have gotten so strong it swarmed in the fall, leaving the new virgin queen faced with a few dying drones who were hungry and not in the mood for love. Thridi might have been soaked by the flood. Or it may have been left in the summer with too many old bees, too few young and strong, a dying population that would not be replenished over the winter, too scant in number to keep themselves warm. It could have been foulbrood that killed them. To learn that, I'd have to wait for a warm, sunny day to take the hive apart and inspect for discolored cells. Or it could have been something else, any number of something elses.

I telephone Red the next morning. "I'm sitting in my office looking at a cloud of bees through my binoculars," he boasted. All his hives were thriving.

"A cloud?" I wanted to see a cloud of my own. I hurried back to the field. Queen Mab did not let me down. The girls were out warming their wings in the sunshine. But here was another wonder: bees were going in and out of Har and Thridi, too. In and out! Was it possible that I'd been wrong? That the hives were not dead?

The sketcher was back to draw the barns. I greeted him as if he were an old dear friend. "You were here for the death," I declared. "Now you're here for the resurrection!"

I couldn't wait to tell the Bee Master. He did not greet the miraculous news of life after death with faith, awe, and jubilation.

"They could be robbers," he proposed. "You want a warm day when they break cluster. Then, when you go in and inspect, first of all you see if there even was a cluster and then, if you can't tell what's going on, well . . . If the bees you're seeing are working their own hive, they're gonna rise up to defend it. Be sure to wear the veil and get the smoker going. If they're robber bees, they'll ignore you. They'll just figure you're a robber too. And by the way, it's six weeks to April, get your Apistan in." Resurrection or rot, it was six weeks to April!

Spring would come as it always came. He told me that the maple, the skunk cabbage, and the elm already had pollen. In mid-February. That was a miracle of another stripe.

I got busy relearning bee pharmacology. Apistan, Terramycin, Fumidil-B, menthol. Rules and recipes and rubber gloves. I knew that for however long I was a beekeeper I would have to relearn the formulas twice a year. It would never stick. But what did I care? It was six weeks to April.

It started to rain and it rained all day. It rained all night and the next morning. There was a rawness to the air that spoke of the current season and not the one upcoming. But I was in on the Bee Master's secret so I bore it without gloom.

Two days before my birthday we began to hear warnings of a coming nor'easter dumping many feet of snow. I purchased two pairs of snowshoes on sale, along with the appropriate outfits, in case the occasion warranted taking up a new sport. We went to the IGA and stocked up on candles, batteries, UHT milk, hearty blizzard food. And birthday treats. From the P.O. box, I collected a gift and a card or two, along with some bills that could wait, and my college alumni magazine.

At the kitchen table, over hot chocolate doused in fresh whipped cream, my daughter delved into Narnia and I into the present of my past. My classmates had become movie stars and mommies and management. The usual things. But someone from the year before, a wiry fellow I remembered from the dining hall, was now the first white chief in the Kwamang Traditional Area of Ghana. If that improbability was fact, then why not dead hives alive?

When a nor'easter is on the way, New Englanders become excitedly tense. They talk of snow as if they'd never seen it before. They talk of snow and nothing else. School and all nonessential services had been canceled in advance of the coming blizzard. When Boston was buried, the news flew down Main Street. The men were ready to lower their plows and start up the salt trucks. They waited. Shopkeepers waited. The superintendent of schools waited. The emergency medical technicians waited, as did the ambulance driver. Parents waited. Children waited. The sun shone white, painting silver linings on pewter clouds. At two in the afternoon, a few flakes fell in delicate spirals, melting on the ground. My birthday blizzard was unlike any other in recent memory. It could be shoveled away with a teaspoon.

My landlady practiced the difficult art of saving money in a most unusual way. Each time she did a load of laundry, if, by hanging her clean wash to dry on a line in the backyard, she then saved the electrical cost of using the dryer, she put a dollar in a jar. March roared in on high velocity winds. Though it was cold enough to start, the wind chill made the temperature feel like twenty degrees below zero. It hurt to go out of doors. High wind and wind chill below zero did not deter my landlady. She stood against the biting air, clothespins in her red, chapped hands, saving dollar after dollar as she clipped the wet clothing to the ropes. Within the hour, whatever clothing was not whipped down to the dirt-dappled snow, therefore needing a second washing, was frozen into planks. She would leave the planks for days, by which time the icy weight would be too much for the clothespins and the frozen clothes joined their fallen comrades on the ground. When, eventually, she returned to pluck her laundry from the line, she would not find it there. She would cluck to her lazy daughters about having to rewash all their clothes, rehanging the same load under similar conditions, putting another dollar in the jar. Eventually, the weather became kinder, at which point I imagine her rate of savings dropped off a cliff.

I had been invited to a black-tie dinner at the home of some people I adore. I was looking forward to a rare evening in the exclusive company of adults. I had a purple velvet gown and my grandmother's purple amethyst pendant to match. Days before, I had scheduled an appointment for a manicure on that afternoon because I had mommy nails, beekeeper nails, dishwasher nails, and I wanted pretty nails, groomed and polished, if only for one night. On the morning of the evening, I woke to a perfect day. The air grew warm as noon approached. It was everything I had been waiting for, a day to go to the bees. I didn't want to waste a minute. I hadn't one to waste.

By 11:30, I was in uniform and at the hives, ready with the Apistan, the Terramycin patties, my hive tool, a patch of hardware cloth, wire cutters, staple gun, a handful of plastic bags and gloves, a bee brush, and an empty hive body. The first thing I did was open Thridi. It was as I had left it. The wandering bees paid me no mind. The rest were frozen in place. With a hearty rap to the side of the lower super, I gave the hoped-for miracle a final chance. No resurrection here. Just stillness. Death is very quiet, even on a sunny day. With my hive tool, I pried off one frame after another, brushing dead bees to the ground. One frame was covered with the carcasses of bees lodged face-first into the cells. This frame told me a sad story. Starvation had driven the bees into their honey stores, breaking cluster then freezing to death, unable to survive without each other's warmth. As I continued to remove the frames, I found the cluster, dead, but still in a ball, between the upper and lower hive bodies. I put a handful of these bees in a plastic bag so the Bee Master could examine them for disease. I pulled the hive apart one frame at a time. This vital nation was now a tomb. The top hive was so honey-heavy I had to place half the frames in my spare box before I could lift it aside to reach the bottom. I cleaned and swept the dead away and then reassembled the hive.

I turned with gratitude to Jafenhar. They were busy at the entrance. The sight of living bees in the winter is akin to a benefaction. What they have weathered in the way they have weathered it, we could not. I did not take apart this hive. I inserted the medicines, Apistan—which is not to be touched by human hands—and Terramycin patties, which have the goopy, sticky consistency of misassembled cookie dough. In exchange, the bees refilled my heart. I crouched beside the bottom board, watching them with love and pride as they went in and out through the tiny portal in the entrance reducer.

I checked my watch. I didn't have time to examine Har. But I did it anyway. Couldn't resist. Because I was pretending that I wasn't really going to do what I was going to do, I did not proceed in an orderly fashion. I lifted off the outer cover. There were bees! Hundreds. And not a one gave a stinger's damn about me. It was carnival time for the marauding ladies of Jafenhar, that's what it was. Plenty for all. As long as their bacchanalian revels could proceed, they didn't mind if I scraped and hacked at propolis and jimmied out the frames. Soon I had the top super off and the hive was officially uninhabited, except, presumably, for the mouse who'd gnawed through the entrance reducer. I did not see signs of death as I had with Thridi. I did not see signs of anything much. It just looked as if they'd all upped and gone.

With the top super out of the way, I began to pry at the lower frames, starting with the outermost frame on the river side of the hive. I wiggled it up out of the box. It was brownish and beeless and it had an arc-shaped quarter of the lower back completely eaten away so that there was an empty space. The mouse! I removed the second frame. Same as the first. Looking down into the hive, I saw that my tenant had shredded field and forest debris so finely that it had become a gray fluffy, puffy down that looked very comfortable. Her nest. I pulled out the third frame and found more damage. She'd gnawed at the honey combs. There was neither honey nor brood. Perhaps she'd eaten both.

In spite of the destruction the fourth and fifth frames disclosed, I admired the lovely home she'd made for herself. A lot of hard work had gone into it. And yet, it would have to go. I assumed that my warm and well-fed mouse was out gallivanting in the sun like the rest of us. I lifted up the entire bottom super and set it on the pale winter grass. I took the bee brush and wiped away her lush, cozy ball of insulation. Onto the ground it went, so light a breeze quickly carried it off. I considered my eviction complete but not satisfying. Such a nice home, and I'd wrecked it. I continued to examine the box. Sixth frame, seventh frame, gnawed. My tenant had lived in the sweetest possible luxury. I stooped to pick up frame number eight and shrieked with surprise.

Looking up at me with exquisite, impenetrably dark brown eyes was an equally startled and extremely frightened little creature. She stared at the giant who had ripped apart her winter palace. I stared back. She was large, quite chubby, and possessed of a most delicate beauty. I felt both ashamed and amused. What a roly-poly deer mouse. She had lived like a baroness in her fine castle, the wealthiest hive of a summer past. If the bees had remained in the hive, they might have broken cluster on the first warm day and stung Her Excellency into permanent exile. But the Fates had woven her destiny from brighter cloth. The bees of Har had left their thick, nutritious honey for her gustatory pleasure and headed off to their own probable demise. She stared. I stared. Then she scaled the back wall of the exposed super. With one leap, she was out. She scampered five or six feet and stopped, turning toward me, toward her lost palace, regarding us both for a long time before deciding to move on. I stood still, my heart torn. My intervention was necessary, from a beekeeper's standpoint, but in that moment it seemed meaninglessly cruel. I could replace the frames. The bees were gone. Come May, I'd replace the bees. Could she replace her home? Having done the deed, I put the hive back as it had been and stapled hardware cloth over her front door as the Bee Master would have me do, all the while seeing her eyes as she looked homeward.

I had forgotten completely about time. Three hours had passed. I had intended to shower and do my hair, to ready myself before my nails were painted. Now I was a half hour late and if I didn't want to be later still, I'd have to arrive in my muddy beesuit to plead my case. When the manicurist saw the state that I was in, pity replaced annoyance and she slid my filthy fingers into fragrant soapy water.

In my tiny household, we do not beware the ides of March. We celebrate it. It is one week before the official start of spring. It is also my daughter's birthday. Turning eight on a school day meant fruit punch and twenty-three chocolate cupcakes, topped in blue with chartreuse icing frogs on dark green icing lily pads. The peepers were not yet peeping, but we had faith they would.

There were tiny pink-tipped crowns, maybe four millimeters high, on the tips of the maple tree branches that dipped behind the hives. The river was deep bottle green. In the field, the dried gourds with their brittle shells looked like faded pottery cast upon the snow by a hastily departing tribe.

It was technically too cold for the bees to be out, but the ladies of Jafenhar had not read the beekeepers' manuals. They were cleaning house, taking out the winter dead. Inside, the queen was laying and her daughters were already tending to the eggs that would become the bees that would gather the nectar in late spring, the bees that would tuck away the treasure to ripen, the bees that would feed the summer bees who would spend their life assuring the hive would be ready to survive the trials of winter but would not, themselves, live to see it. The springtime bees would know the meaning of a chilly night, they might encounter disease or predators, but only the hearty queen would be alive on the earth long enough to travel one full revolution around the life-giving sun.

I visited the Bee Master with remains from Thridi and he confirmed the verdict. Starvation. He examined the two frames I brought. The frame from the lower super was riddled with drone cells, no workers. The queen had been laying in a strange way before her demise. If she were ailing, the morale of the hive would have dwindled. In winter, there would be no hope of continuing their line. Har's fate was more mysterious. A swarm, yes. After that, a weak queen, yes. But why was there so little evidence of her brief reign? There was no sign of disease. I purchased a new beekeeping textbook for $29.99. I treated myself to ten preassembled DURAGILT frames and a new entrance reducer to replace the mouse-gnawed one at Har. The Bee Master gave me excellent reasons why I should have three slatted boards to place above the bottom boards. These needed complicated putting together. I bought them anyway and promptly forgot why they were such a good idea. I purchased some pale yellow pollen substitute for my happy hive. With this, and a new recipe, I could concoct a helpful springtime snack.

"I can't believe you still got that dinky Honda," the Bee Master teased. "When you ever gonna get around to realizin' that a girl like you needs a truck?"

A girl like me . . . I used to collect Hermès scarves.

That afternoon, my daughter and I gathered tiny pinecones in a basket. The sky was icing blue and the clouds were cotton candy, you could taste the new season in the air, and everything seemed possible. I left her whirling on the lawn. I mixed poisonous menthol with canola oil and saturated paper towels. That would help deal with mites. The pollen substitute was a powder requiring the addition of two parts sugar to one part water. In the end, it looked so temptingly similar to chocolate chip cookie dough before adding chips that the only thing for it was to hustle my daughter into the Honda and dash to the IGA for brown sugar and Nestlé's morsels. The bee dough was rolled between waxed paper pieces and cut into squares the size of candy bars to be placed on top of the frames. The cookie dough was hastily, happily devoured instead of dinner.

The next day, my daughter saw a young goldfinch splashed with yellow. I noticed a bluebird that looked almost purple and a crow carrying a twig into the forest for its nest. Six violet crocuses had opened beside the driveway. And my bees were bringing the real thing, pale butter-yellow pollen from some newly opened something somewhere, into the hive. Spring had parted her veil.

I received a beekeeping newsletter in the mail. It issued a harsh directive about quotas for queens. If the old girl wasn't laying at maximum capacity, they advised killing and replacing her. I paid no attention to that. It came from Type A country, a frenzied land downstate filled with stock market winners commanding carnivorous cars and furnishing tract castles. The Bee Master was my teacher. He trusted the bees to make their own decisions as they always had. He believed we were there to help and to harvest, not to hire and fire.

The second day of spring it snowed, but there was no going back. There never is. Though the colors around us were still winter-muted, we woke to choruses of ebullient birdsong. One day the river was racing, the next day it hardly appeared to move. Within a week, the air was filled with the roar of saws and weed whackers. I fled to the field seeking the company of my bees, anticipating the quiet whispering of the river and the splash of rising trout. Instead, I found several trucks with the name of a local landscaping firm painted on the side and a vehicle that resembled a bulletproof golf cart with a giant yellow scoop on the front, which was operating at full growl. A large white Dumpster stood between the barns. It was already nearly filled with brush. There were three people clearing wood off the spit of land dividing the river from the small tidal pool. The wetland weeds had been yanked. A giant maple, four feet wide at the base, had already been felled and sliced into large chunks. Someone else was pulling up roots at the water's edge. The guy operating the scoop had earplugs in his ears. I wandered over to a knot of men in sweatshirts and hollered the only possible question: "What in heaven's name is going on?"

"We're mulching the footpath."

"Footpath!" I cried, "What footpath?"

"The one they're puttin' in."

They. I knew who They had to be. So. They, the they that is never the we that we are, were at it again. "Where?"

He pointed to the narrow spit. "Out around there and along the river."

Damn. I'd heard rumors They'd been thinking of putting in a petting zoo and laughed them off. This was nearly as bad. The scoop man had stopped his scooper and taken out his earplugs. We no longer needed to shout. "What do you think about that?" The standing-around men looked at the scoop man. The scoop man shrugged. I gestured toward the hives. "You know, I'm out here all the time. I've seen herons, turtles, muskrats, all living right out in the water and on the spit. Some of them lived in those roots. That's their habitat."

"I seen beaver, too," the scoop man admitted. "Swimming around."

"If you pull all that out and people start walking around out there—"

He heard the plea in my voice. "I know. You know. But They don't care about nature! They just care about property values."

A good part of the economy was structured feudally. The particular They to whom we all referred wrapped themselves in land and cash and paraded northward. They were the lords. Locals, particularly tru locals, didn't have that kind of money and never would. Sometimes they farmed. Usually they served, built, fixed, pampered. The locals needed Them. But need is no guarantee of respect.

We were all of the same mind about Them. Our opinion didn't count. They controlled this land, were charged with its stewardship. If the farm was farmed, it was done at their sufferance. My bees were guests. My fellow disgustees were local laborers. They could not afford to risk their jobs bucking Them for the sake of good sense and some muskrats. Laying down a mulched footpath for those who preferred to stroll through a muddy field by a riverside without soiling their loafers would mean half a week's wages.

"We'll watch out for your bees, too."

My bees. The mulched path would lead straight to the hives. Even after the flood, I had given little thought to moving the hives. I didn't want to think about it. The prospect seemed so daunting. I didn't know how. If I let my mind wander in that direction, I entertained half fancies that a simple move to higher ground, nearer to the road, might suffice, that I wouldn't have to leave altogether. That dream was over. The alarm clock was ringing. May first was bee day. With one live lively hive and twenty-four thousand new bees on the way, where would I go? What would I do? I felt like Scarlett O'Hara without tomorrow. I had to get them off this land. Fast. Yes, there was the flood plain. But Their footpath was a more pressing good-bad reason to go. If a tourist ambling along a newly sculpted public footpath situated beside my hives got stung by an aggressive wasp, went into anaphylactic shock and died, it would not be up to the family lawyers to prove the cause of death. Three beehives stood right there on the property. Bees stung. Some people were allergic. Everyone knew that. The accusers would trot out their Latin. *"Res ipsa loquitor."* "The thing speaks for itself." Case closed. It would just be a matter of assigning culpability. My fault. The Land Trust's. Or both.

"You be sure and do that," I said.

A year before, I had barely been aware of these eleven acres. We were strangers thrown together by happenstance under some white plaster elephants. Since that time, I had walked this land in all four seasons, lived with it, worked on it, learned from it, loved it. I loved the creatures that lived there. The river. The sounds. It was the place I hurried to in search of joy or comfort. And the bees—they were part of my family now. I needed them more than they needed me. I picked up the head of a discarded old pitchfork. Its rusted prongs suddenly seemed very, very precious to me. I put it in my car and sped away. I didn't want to cry in front of all those men.

How was I going to find new land? I couldn't afford to buy my own. The farmers I know grew cow corn.

It was Pete the crossing guard, Pete who ran the dump, Pete the fix-it man, Pete who rode around town in his truck with his petite, well-coiffed wife in the front seat, his pumpkins overflowing the bed and the trailer behind, Pumpkin Pete, who knew whom I should turn to. He handed me a sheet of paper torn from a notepad. It said PipeCo on the top. On it, Pumpkin Pete had written "HOME FOR YOUR BEES!" with a name, an address, a phone number. I taped the paper up on my window frame just under my motto, "The unlived life is not worth examining," and looked at it anxiously for two days before I gathered the nerve to leap the abyss, shift my shape, fly into the air like a bird off a branch and pick up the telephone to make a local call. I didn't get Mr. Sweete. I got Mrs. Sweete, a retired Yankee schoolmarm, who wasted no time terrifying me. I explained my purpose in calling. "Call back at six and speak to him," she answered curtly. I obeyed. He wasn't there. I agreed to telephone once more around one o'clock Sunday after church. I telephoned at twenty past, thinking I'd allow time for Sunday supper to settle. "You said you'd be here at one," she snapped. "He's out the barn waitin' for ya."

"Oh, dear," I sputtered. "There must have been a misunderstanding."

Mrs. Sweete crackled. "Well, tell it to him."

Afraid I'd made a mess of things before I'd even started, I sped like a demon to the barn. All the way I was cursing myself. I arrived prepared for a dressing-down and found a smiling eighty-year-old man, his face a map drawn with rivers and streams, loading fifty-pound sacks of feed onto his truck with the ease of a child casting petals on a pond. I introduced myself with a string of apologies. "Aw, that's all right," he said, and I knew it was. He walked me across the road to the spot he favored. It was in a corner of a clearing across from his red barn, a few feet away from another made of cement blocks. Mr. Sweete had had lots of cows and the whole place smelled of manure.

On the bright side, it was a place. It was near to the road without being too visible. Sunny, but not too. Near a pond so the girls could sip. No threatening riverbanks. Rather sheltered from the north winds. Very sheltered from the west. And it was most certainly on a farm, for in addition to growing cows and corn, Mr. Sweete had fields of sweet clover and alfalfa to supplement the wildflowers. Bee food aplenty. When one moves bees, one wants to be sure the move is far enough, otherwise the bees will try to return to their usual home and find it gone. Mr. Sweete's farm was a bit more than three miles away. Perfect. I could be confident that after some initial confusion, the ladies of Jafenhar would quickly reorient and adjust. I was not convinced I would do likewise.

I visited and revisited the place, pacing the little plot of land. I came in the afternoon. I came in the morning. I brought my daughter for a look-see. I consulted with the Bee Master. They told me it would be just fine.

Mr. Sweete waved every time he drove past in pursuit of his endless farm chores. One afternoon, on his way home, he pulled over in his blue truck to tell me he was happy I was bringing my bees. He'd been remembering how his father had had bees when Mr. Sweete was a boy my daughter's age. "I love honey," he said. I promised him lots. He told me his wife was eighty-four.

"You married an older woman."

"Yup. She chased me and chased me 'til I gave in."

Mr. Sweete's daughter, grandson, and granddaughter were all volunteer firefighters. Born here. Stayed here. Gave to the community. Mr. Sweete's granddaughter was also my daughter's second-grade teacher, who taught in the same classroom that had once been ruled by her no-nonsense grandma. Mr. Sweete rose every morning to rouse the dawn with the roar of his old tractor. If I didn't love the land straight off, I began to trust that I would. I would love the land because it was his land and he was irresistible.

I abandoned my hesitation. This changed the reason for my visits to the land. I needed to figure out how to place Har, Jafenhar, and Thridi so they would rest level and face the rising sun. The smell of dung seemed to bother me less. And when the wind changed direction, the smell of dung disappeared.

I would move the ladies of Jafenhar and my two empty hives. My new live bees and their new live queens would come on a flatbed truck from somewhere in Georgia. They would live, separated by a rose-covered fence that was not without its thorns, beside Mr. Sweete's cows. We would have milk and honey.

10

What Changes, What Remains the Same

Take the honey with the milk,
drink of it before the rising of the sun,
and there shall be something in thy heart
that is divine.

—*from the Berlin Magical Papyrus*

WAS SPRING. NOT ALL THE WAY, BUT ON THE WAY. My girls of all species had discovered daffodils. My daughter picked them. Athena tried to nibble them. And the merry bees of Jafenhar had not only the baskets on their back legs filled to the brim, they were covered head to toe as though they'd been rolling in sunshine dust.

Red offered to move the hives to Mr. Sweete's farm on the back of his four-wheel-drive truck. We agreed on a time and a plan.

The day before I moved my bees, I was as nervous as if I myself were being uprooted. I went to the hives to remove the last bits of Jafenhar's medication and found the stone that had been sitting atop Har for nearly a year had been cast to the ground. The stone had weathered storm and flood. Only a human hand could have removed it. Someone had already tried to poke their nose under the covers. Fortunately, they'd picked an empty hive. No matter how I loved this patch of land, it was clearly time to leave it. The Bee Master told me to purchase thick nylon straps with ratcheted clasps. I was to wrap the hive top-to-bottom and back again, binding Jafenhar's two hive bodies together. But the straps from the True Value were way too short to fit around a couple of stacked deep supers. I was then to join the bodies together by nailing a strip of wood like a diagonal bridge over two sides. Compensating for the missing straps, I nailed all four. To insure that all were ashore that were going ashore, I waited until dusk to plug up the portholes and staple a screen over the mouth of the hive so that the bees would not fly out the bottom. The dead hives were a cinch. Duct tape did every job that needed doing.

On April 15th, we rose before dawn without a thought to taxes. Because bees are bees, I could not wear my heart-shaped diamond ring or my linden flower perfume in recognition of the importance of this occasion. Yet the day wanted marking in some personal way so I donned my fanciest underwear. Over that went more mundane layers and my clean Cloroxed beesuit. My daughter wore two sets of sweats and farm-boy coveralls. We ate our first breakfast of the day.

There was frost on the Honda's windows. The cold was useful. It meant the bees would stay in their hives. That gave us a smidge more time. Red arrived just after we did. He drove his truck straight across the field and backed up to the hives. Bending at the knees, one of us on either side, we lifted the honey-laden Thridi first. Har was lighter. The mouse had seen to that. The last to load was Jafenhar. Dawn painted the clouds to match my daughter's flushed cheeks. Helios mounted his chariot to pull the sun across the sky. But he had yet to show himself. Chill and darkness were still on our side. Jafenhar had to be lifted level and all of a piece. We raised the hive onto the truck in a swift smooth arc and then wiggled it back into the bed so it was cradled by the others. Red wedged my worktable against it so that it would not tip. In went the cement blocks that held up the frames. In went the frames on which the hives stood. Then away without a second for bittersweet adieus. First three miles south, then east.

We arranged the blocks and frames in a staggered triangle. Har was nearest the rose-covered fence. Thridi nearest the road. Jafenhar resumed center place without incident. All three faced east toward the rising sun. We heard the call of a ring-necked pheasant. The air was only slightly warmer. We were hot from lifting but didn't care. Greeting the morning as we had, what could a small discomfort matter? I removed the screen and unblocked the portholes. The girls would be free to explore.

All three of us clambered into the front seat of Red's truck. He drove down the road another mile east and then turned off over rocks, a fallen plank bridge, a shallow brook. The truck negotiated a steep and deeply rain-rutted hill as if it were a heavy-duty tank. Red guided it up and up into a high field. He cut through the field and the truck climbed still higher until we reached a hilltop pasture. "Here's where your bees will soon find their clover," he said. The three of us sat as silent as the growing clover looking out at the world, the hills newly greening, the buds just barely showing their tips, hills all around, gently bosomy hills, no edges, just folds, no sound but the pheasant's cries. Reluctant to turn away and allow time to start again, we ventured downward over the jarring bumps toward the diner in town. The ladies of Jafenhar were already tentatively testing the air. Red stopped at the old field long enough for me to retrieve my car. That was all of me that was left there. The landscape felt empty without my hives. Plastic bottles and potato chip bags already littered the mulched path. I followed Red into town.

At the local diner, Red ordered oatmeal, three scrambled eggs, sausage, and toast. My daughter and I each downed a stack of pancakes, eggs-over and bacon. We gloried in the aspects of this morning that had already become memory until my daughter grew anxious. She was in a hurry to get to school in time for recess so she could recount her adventures to her friends. Red headed for work in his powerful truck. Dreamy and idle, I drifted through the rest of the day like a pampered drone. So what about laundry? So what about food? So what about work? We had moved the hives away from their lovely wrong place and onto their smelly right one. My girls had a fine new home where the sun would bless them and the trees shelter them from wind. All this, and two good breakfasts in one morning. What else could possibly be worth bothering about?

First deep, early purples, then the brightest yellows, radiant greens, uncountable pinks, whites in all shades. Let the rainbow color neatly within the lines, spring splashed about higglety pigglety. The oak in my driveway sported a pine tree growing in its crotch, ten feet above the ground. Across the street from the second-story pine a flowering tree in the gun-toting sociopath's front yard sported one branch with white flowers and one branch with bright pink. My daughter spied a large, smooth, gray boulder under another pink-flowered tree. The wind was blowing. It had rained in early morning. Many petals had fallen into the meandering crevasses in the rock, making a gentle pink aerial map of branching tributaries. Little brown maple whirligigs flew through the sky, falling like parachutists behind the lines into the green grass to seek cover. At twilight, the dogwood blossoms absorbed the last of the sunlight and glowed white even into the night.

As our second Bee Day approached, I was overcome with a resurgent case of acute hammerphobia. The prospect of nailing together ten DURAGILT frames caused me to freeze. I could either fight it or give in.

I gave in, and traveled twenty-some miles south just to avoid this simple task by purchasing them ready-made at an outrageous price.

The Bee Master received me warmly. "You remember that talky woman who was in here?" I did. Vaguely. "Well, she come in the other day asking for your number. She said she wanted to help you. She's had her hives for three years and she's yet to get honey out of those bees. No way I was going to give her your number. I don't want her giving you her 'help'." And then he spoke words I had never hoped to hear. "You're a much better beekeeper than she is!"

I? A better beekeeper? Than someone? Anyone? I chortled in my joy.

The difference between installing the first thirty-six thousand new bees and the second twenty-four was that I did not need to buy a saw at the last minute. I knew what to expect. I sawed the wooden cages apart, squirted calming sugar water on the girls and listened as their tone changed, gave each cage in turn a hard knock upon the ground, lodged the queen cages between the fourth and fifth frames, removed half the frames that remained and shook the two new populations, one at a time, over their uncrowned virgin queens and into their new used homes. Unlike the inhabitants that worked these supers before them, my new girls would be able to dine on honey straightaway. They would benefit from the bounty left behind by their predecessors. These new ladies of Har and Thridi would also have the advantage of drawn-out comb. Their work would be easier. I lingered like a lover under the sparkling powder blue sky. My hives were full. So was my heart. The gratitude I felt at finding myself in the service of these unknowable companions whose earth was my earth, whose world was unlike my own, was too deep for words.

In Southern Maine, a truck carrying a cargo of roughly twenty million bees overturned on a highway ramp but no one was hurt. The farmers had no more fear of frost. Corn and flowers could be planted. Demeter could dry her tears and embrace her daughter as she stepped once more into the light. My daughter gave me handfuls of pipe-cleaner hearts and notes which all said the same thing: I love you. It was Mother's Day. I counted myself among the luckiest of women.

Wearing my fancy French reading glasses, I inspected the homes of my new girls for day-old eggs to see if newly installed queens had taken their mating flights and settled in, like me, for a lifetime of motherhood. The new queen of Har was long and tawny gold. I was lucky to see her. Lucky twice, I watched the new queen of Thridi walk across a comb I lifted from the hive. Like the ruler of the neighboring principality, she had a black head and black thorax, and a black tip at the end of her pale striped body, which reminded me of armadillo armor, only smooth. The queen of Thridi chose a cell and stuck her head in. Finding it to her liking, she then backed herself bottom-down into that same cell to lay her egg. She did not lay assembly-line style. She wandered a bit more, selected the next good place, laid another egg.

I had not expected to see Mr. Sweete driving anything other than a truck or a tractor, yet there he was in a family car. He looked so out of place it took me a moment to recognize him when he pulled over. Mrs. Sweete sat by his side. They were on their way to lunch. I would never have taken her for eighty-four. She was slim, a beautiful woman, wearing bright red lipstick and a delicately flowered blouse.

While poking around Har, I had placed an outer frame in the box. It lodged in an awkward position. When I tried to dislodge it, it fell hard into place and there was a crying out, what seventeenth-century bee-keepers would have called a "trumpeting." A small but opinionated group of protesters whirled about my head, delivering a firm warning. No sting. Just a rebuke. I had broken off a chunk of honeycomb and wrapped it in paper towel intending it for my daughter. Instead, I handed the comb to my new host, Mr. Sweete, who popped the whole thing, wax and all, into his mouth in one great happy gulp and chewed. "You sound like you have false teeth!" Mrs. Sweete scolded. But she also laughed. I told Mrs. Sweete that the queen bee was a single mother who laid between a thousand and fifteen hundred eggs a day. "Got her hands full." Now we all laughed. I was getting the hang of Mrs. Sweete.

Monday, Tuesday, Wednesday, I delivered supplementary sugar water to Har and Thridi. Jafenhar was well beyond a fast-food diet. There was nowhere to rest a finger without first moving a bee. On Thursday, I put my first springtime honey super atop last summer's struggling hive. The earth had gone once around the sun. The weak were strong. The strong had gone.

I could already see that the new queen of Thridi had an edge. Her girls were defensive, ready to sting. When I approached, they went for the head and wasted their lives on my gloves. Har was a small, peaceable kingdom. Everything had changed except the names I had given them.

On Friday, I had no time for bees. I took my daughter to school and hurried home without even stopping to buy the newspaper. I had endless chores to finish before the weekend. Dishes. Laundry. Bills arrayed on the table. I put water on to boil for tea. The kettle whistled. The telephone rang. It was Red calling from his office.

"You gotta come see this!" he shouted. "A big black cloud of bees just flew out of my hive. They hung there for a moment. It looked like all of them at once. And then it seemed like they were deciding who was going and who wasn't and about half of them took off. A big black moving cloud. And they settled on a tree. You gotta come. I got a big oil drum and I put a swarm trap on top of it. You gotta come."

What other people call real life would have to wait. This was realer. "I'll be right there."

"Do me a favor?"

"Sure."

"Stop by my house and get my beesuit, my yellow brush, and my gloves." I grabbed my beesuit and a pair of socks and hurried out the door. In order to get into Red's garage, I'd have to break, enter, and disable the alarm. Leaving the motor running, I committed my crime, clambered over bikes and toy Jeeps to the beekeeping side of his garage and found his suit in a snap. The road seemed to stretch under the wheels of my blue Honda as it lumbered along at fifty miles an hour. I couldn't bear the thought of missing this swarm. This queen, the monarch moving out and on, was the very first queen I'd ever seen on my very first day as a beekeeper. If I couldn't catch her, at least I wanted to see her off.

As I pulled into the parking lot, I looked up and around. No bees. When we were both hatted, veiled, zippered, taped, and gloved, we crossed the lawn together. He pointed. "There."

Bees hugged the slim trunk of a tree as if they were a thick, undulating extension of the bark. And I hadn't noticed. Bees stood upon the backs of bees, linking legs, wandering, fanning their wings. The queen was invisible, buried and protected under the shield of her many thousand children.

"My smoker's ready if you need it."

"They won't sting. They're engorged with honey."

We began. Standing on either side of the tree we positioned the trap. As we brushed, the bees fell in thick clumps. Just where we wanted them. Bees whirled around our heads and hands as we worked. Enclosed within their living cloud, enveloped by the profoundly stirring music of a single vibrating note, we cleared the tree, or most of it, sweeping the bees into Red's swarm trap. Fear kept well away. To be in the midst of this swarm one had to be serene. And we were. Peace came without intent or effort. Red slid the lid over the teeming masses. I placed the trap on the ground with the entrance facing the tree. If the queen was in the box, they would join her. Some of the bees returned to the tree trunk. Some clung to the outside of the box.

Red made the call. "I think the queen's in there. They're trying to get in."

The lid was loose, but I wasn't sure it was loose enough. I took a long stick and gently lifted the bees into the trap. They clung together and on the stick so that it looked like a cattail weed. But in they went. Red stood back. Crouched beside the box, I took the stick and made a gap between lid and trap just big enough for a bee to enter.

I realized that as I held the bees captive, so could I also set them free. If they were in the wild, they'd be facing an environment that has made survival nearly impossible. If Red tended them and guarded them from enemies microbial to mammal, they had a better chance to live and prosper.

On the way home, I decided to drive past Har, Jafenhar, and Thridi. I was already in my beesuit. Why not take a quick peek before getting down to all those abandoned chores? I checked the honey super first. Not much action there. Four bees wandered the combs in a desultory way. Odd. Where were the others? They'd been so anxious to climb up through the queen excluder just the day before. Double odd. I decided to go through the hive. The combs were not as crowded as before. I found one capped queen cell hanging off the bottom of an inner frame. Then another. Then another. Seven. Eight. Queen Mab had decided to start another nation, had she?

No! I thought, You can't go!

I began to remove the cells, setting them down on my work table. I'd just happened to be reading, though I didn't remember where, that this was the way to forestall a swarm. Buying time—as if it could be bought. I took a bit of burr comb that might block passage between the supers, and put the lot in a discarded plastic bag that happened to be on my backseat.

I called the Bee Master and told him how I'd helped Red capture his swarm and temporarily staved off one of my own.

"You removed the *cells*? What the hell'd you do *that* for?!" he yelled. "Were any of those cells *capped*?" I answered meekly. "Well then, you've probably gone and made your hive *queenless*!"

"I thought I was supposed to—"

"You *thought*! What'd you do? Look in a *book*? What *they* don't tell you is that sometimes the old queen'll leave the hive before the new one emerges!"

"What do I do now?"

"You wait. And you keep checking the hive for eggs. In two weeks you'll know. And if you got no queen, you're going to have to come down and I'll get you a new queen and you're going to learn how to requeen your hive. *That's* what you're going to do!"

While I was being roundly chastised for multiple misdeeds and misdemeanors, I wandered into the kitchen. I opened the plastic bag and emptied the contents onto a flowered paper plate. I consoled myself with the thought that since I had made a thorough mess of things, at least I could show my daughter the queen cells. "What you *should* have done . . . " I noticed that three little bees were wiggling their way out of the burr comb. And then, exactly then, a young queen emerged from one of the cells I'd so hastily, foolishly removed.

"Um—" I interrupted cautiously, "a queen just emerged in my kitchen. It's definitely a queen. Long body. Short wings." Athena's tail had begun to twitch with excitement. I swiftly carried the plate outside to the porch.

"Take her right straight back and stick her on the bottom board and let her walk in. Then pray."

"OK."

The queen shot straight up into the air and hovered just out of my reach as if she knew she ought to do something but didn't know what it was. We were in the same predicament.

"I gotta catch her."

"*Newborn queens don't fly!*"

"Well, this one does."

I rushed back into the kitchen and grabbed an old Stonyfield yogurt container. It smelled faintly of French Vanilla. Maybe that would help.

I poked holes in the lid with a barbecue skewer, describing my moves to the Bee Master.

"Don't worry about that," he said with the slightest trace of patience returning to his voice. "Just be gentle. They're real delicate. Remember, she just got born."

This time, I had honey to hand, our honey. I dropped a glop into the container and tore a sprig of thyme off a potted plant. Thyme. Two thousand years ago, the Greeks thought highly of thyme. So did bees. Why not now? I set about trying to coax the infant queen toward the anointed yogurt container. What about the burr comb? Would she be drawn to that? In it went. The hovering queen approached. She touched the edge. She retreated into the air. I tried to trap her with the lid. I tried holding the yogurt container aloft. I didn't want to drive her away. Gentle. I returned to the kitchen. By the washing machine, I kept nylon net bags for stockings and lingerie. I opened one wide and scuttled back. What if she vanished? She had. But just for an instant. Then she returned, lighting on my rusted barbecue. Careful not to cast a shadow, I swept the lingerie bag over the queen. I brought the bag to the yogurt container and held it so she would fly down toward comb, honey, and thyme. Down went the lid. I heard the wax shift. What if I was squishing her? I lifted the lid, but not too high. She was alive.

"I've got her," I sighed.

"Now *do* something about it!"

I sped to Mr. Sweete's farm. I stuck every single cell I'd stolen back on the bottom board and released the new queen into Jafenhar. If the hive survived my hubris, I'd never know for sure that it was this same queen who ruled. Another, from a cell I might have missed, could have emerged when she did. If so, the pretenders would fight until one was dead and the other ascended the throne.

But if the lingerie bag queen triumphed, what then? Perhaps her strange beginning would have an ill effect. Perhaps she would thrive.

I could not know her fate any more than I could know what life would bring my beloved daughter.

Or me.

It was yet another awkward, bumbling, beautiful beginning in not-quite-paradise.

This much I could be sure of: In two weeks' time, I could hopefully perch my frilly French glasses on my nose, get into a space suit, throw a plastic helmet on my head, and check for day-old eggs.

The Bee Master was waiting for my call.

"So," he chuckled, "you been beekeeping for a year and you thought you *knew* somethin'."

I understood his lesson. No matter how many seasons I lived on this earth I would never become an expert, in bees or in life. And that was a blessing. It would keep me curious. It would keep me humble.

Endnotes

Epigraph

Page vii, "The sting of the bee and the dart of cupid . . ." Lyndy Abraham, *A Dictionary of Alchemical Imagery* (Cambridge University Press, Cambridge: 1998), p. 20.

1. To Just Be

Page 1, "The hive is sweet . . ." Quoted by Hilda Ransome in *The Sacred Bee in Ancient Times and Folklore* (Houghton Mifflin Company, Boston and New York: 1937), p. 74.

2. Spring Ahead

Page 7, "He must be a stupid . . ." *Mackenzie's 5000 Receipts*, 1829, as quoted in *Granny's Honey and Beeswax Prescriptions* (no publisher or date) from Mann Library, Cornell University, p. 59. I suspect this bit of wisdom may not originate with *Mackenzie's*, which may, in turn, be quoting another source.

3. Bee Days

Page 27, "Anyone starting . . ." Quoted in Ransome, *The Sacred Bee*, p. 170.

Page 53, "divid[ing] a . . ." Ivars Peterson, "The Honeycomb Conjecture: Proving mathematically that honeybee constructors are on the right track." *Science News* 156 (July 24, 1999), p. 60.

4. Flowering

Page 61, "When you are stung . . ." Charles Butler, *The Feminine Monarchie: or the Historie of Bees* (Printed by John Haviland for Roger Jackson, and are to be sold at this shop in Fleet Street over against the Conduit, 1623). "29. When any is stung"

Page 65, "sugar". Think also of sugar's relationship to the slave trade in the West Indies. If we'd stuck to honey and kept our reverence for bees, who knows how history might have gone?

Page 65, "In our time . . ." All of the scientific quotations in this paragraph are from Root, Morse, and Flottum, *ABC and XYZ of Bee Culture*, 40th ed. (A. I. Root Company, Medina, Ohio: 1990).

5. High Summer

Page 79, "I am a little . . ." "I Am a Little Italian Bee," words and music by Mrs. J. M. Morgan (New York: Marcellus, 1918).

Page 82, "an admonishment . . ." *The Arabian Nights*, retold by Brian Alderson (New York: Morrow Junior Books, 1995).

Page 86, "Morphic fields . . ." Rupert Sheldrake, *The Presence of the Past: Morphic Resonance and the Habits of Nature* (Rochester, VT: Park Street Press, 1995) p. xvii–xviii.

Page 87, "Theatre of . . ." Samuel Purchase, Master of Arts and Pastor of Sutton in Essex, *A Theatre of Politicall Flying-Insects wherein Especially the Nature, the Worth, the Work, the Wonder, and the manner of Right-ordering of the BEE, is Discovered and Described*

(London, Printed by R. I. For Thomas Parkhurst, to be sold at his shop, at the three Crowns in Cheapside, over against the Great Conduit, 1657).

Page 87, "The males heere . . ." Charles Butler, *The Feminine Monarchie: or The Historie of Bees shewing Their admirable Nature, and Properties, Their Generation, and Colonies, Their Government, Loyaltie, Art, Industrie, Enemies, Warres, Magnanimitie, & C . . .,* Written out of Experience by Charles Butler, Mag: (London, Printed by John Haviland for Roger Jackson and are to be sold at his Shop in Fleet Street, over against the Conduit, 1623). This quote is from the preface, and the quote to follow, regarding drones, is from chapter 3. Charles Butler, a brave and honest soul, is believed to be the first writer to claim, definitively, in print that the hive was headed by a Queen and not a King.

Page 88, While we are on the subject of drones and their behavior, I want to share a little ditty passed on to me by a source who would rather remain anonymous. It is sung to the tune of "Three Jolly Coachmen."

"The Queen bee is a busy soul / She has no time for birth control / That's why in times like these / That's why in times like these / That's why in times like these / We see so many sons of B's."

Page 110, "The distance is indicated . . ." A. I. Root pamphlet "The Language of Bees," quoting *Bees, Their Vision, Chemical Senses, and Language,* by Karl von Frisch (Ithaca, NY: Cornell University Press, 1950).

Page 110, "The dancer's wings, beating at 250 cycles per second . . ." Guy Murchie, *The Seven Mysteries of Life, an Exploration of Science and Philosophy* (New York: Mariner Books, 1999) p. 248. The observations regarding bee dialects are also from Murchie, p. 249–50.

Page 110, "calibration . . ." "Bees log flight distances, train with maps," *Science News*, 157 (February 5, 2000), p. 87. Report on the research of Mandyan V. Srinivasan of the Australian National University in Canberra and his colleagues in Canberra and at the University of Wurzburg in Germany.

Page 111, "There are more described species of bees . . ." Bees take an astonishing lead. The rest total up at 22,600 species. Christopher O'Toole and Anthony Raw, *Bees of the World* (London: Blandford, 1993), p. 32.

6. A Sweet Harvest

Page 119, "The oldest honey . . ." Eva Crane, *A Book of Honey* (Oxford: Oxford University Press, 1980), p. 42.

Page 122, "creamy." Nobody uses this anymore, but it's Marilyn Monroe's favorite adjective in the film *How To Marry A Millionaire* and due to be reinstated in the popular lexicon.

Page 122, "horrible mites and microbes . . ." Varroa mites, Tracheal mites, and Chalkbrood have been in the United States since 1972. Since the late 1990s, the hive beetle has emerged as a new threat to an ancient species.

Page 145, "Figuring an average flight . . ." I have seen this astonishing statistic in many places including *City of the Bees* (Medina, Ohio: A. I. Root Co., 1988).

7. Fall

Page 147, "There's a whisper . . ." Rudyard Kipling, *The Long Trail.*

8. The Chill of Winter

Page 163, "Yes it's a very . . ." The White Knight. Lewis Carroll, *Alice's Adventure in Wonderland and Through the Looking-Glass* (New York: Signet Classics, New American Library, 1960) p. 208.

9. A Mulched Footpath

Page 181, "There is no room . . ." Virgil, *The Georgics*, Book 4, lines 226–227, trans. L. P. Wilkinson (New York: Penguin Books, 1982) p. 131-32.

10. What Changes, What Remains the Same

Page 203, "Take the honey . . ." The Berlin Magical Papyrus is quoted in Ransome, *The Sacred Bee,* p. 278. The footnote reads: *Berliner Zauberpapyrus*, I, 20 (Abt. Berl. Akad., 186, s. 120). It is not entirely clear, but from the context one has the impression the worshippers so instructed were believers in a "mystery religion," perhaps the worship of Mithra.

Acknowledgments

THERE WOULD BE NO *BEEING* WITHOUT PATIENCE, my daughter, and the bees. Its present form is owed to the efforts of Lisa Bankoff, Patrick Price, James Gregorio, and John Frederick Walker, all of whom know well what it took. I thank Dominic and Mary Frances Gaeta with the deepest appreciation and affection. I thank Paul Garlasco likewise, for his energy and assistance. The resources of the Mann Library at Cornell University and the generous resourcefulness of Angi Faiks, Marty Schlaback of the Entomology Library, and the Special Collection librarians, Stephanie, Marjorie, and Eveline were without parallel. I thank Anne and Terry Vance, Sherri Maxwell, Connie Kaiserman, Scott Bricher, Emily Dwight, Joanna Knatchbull, Dana Chevalier, Len Kaczmarek, the wonderful Mr. and Mrs. Newton, charming Brother Daniel of Buckfast Abbey, and the hospitable beekeepers of Copenhagen, each of whose very particular contributions are of great value to me. I thank Audrey and Alec Thomas in more ways than I can list. And last, I respectfully and gratefully thank all the "characters" in my book, all of whose names have been changed to protect their privacy.

Further Reading

*I*F YOU ARE NOT ALREADY A BEEKEEPER AND ARE inspired to become one, the best thing to do is find a long-experienced beekeeper to help you. I was blessed by good luck. There are more methodical ways. Check for local beekeeping organizations in your area. There will almost certainly be someone amongst the membership who will help you.

My favorite basic beekeeping resource is *The Beekeeper's Handbook*, third edition, by Diana Sammataro and Alphonse Avitable, Cornell University Press, New York and London, 1998. I recommend it for anyone starting out.

Another book well worth having on hand as you progress is *ABC and XYZ of Bee Culture*, which is "An encyclopedia pertaining to the scientific and practical culture of honey bees." The original work was by Amos Ives Root. I have the 40th edition, which was written and edited by the late and venerable Dr. Roger A. Morse, with Associate Editor Kim Flottum. It is published by The A. I. Root Company in Medina, Ohio.

Because I am interested not only in the scientific study or "management" of bees—both of which are enormous subjects—but most especially in the long and intimate relationship of bees to man and vice versa,

I recommend most heartily any book written by Eva Crane on this topic. Her *The World History of Beekeeping and Honey Hunting* (Gerald Duckworth and Co. Ltd., London, 1999) is a masterpiece, and as far as I know, the definitive book on the subject.

There are two "children's books" I would like to recommend for both adults and children who want to know more without getting into heavy reading: *Life of the Honeybee*, by Heiderose and Andreas Fischer-Nagel (Carolrhoda Books, Inc., Minneapolis, 1986) and *The Life and Times of the Honeybee* by Charles Micucci (Houghton Mifflin, New York, 1995).

I also recommend any of the works mentioned in the Endnotes as delightful reading.

Bees and the keeping of bees are subjects that shoot out in several directions like the rays of the sun. There are thousands of books, articles, and websites from which to choose, and each book mentions other useful books in the bibliography. You could spend several lifetimes reading about bees, but I would ask that you take at least a moment to appreciate them in the open air the next time you spy a bee at work in a flower.